AF357833

Solitary Fibrous Tumor

Solitary Fibrous Tumor

Guest Editor

Bahil Ghanim

Basel • Beijing • Wuhan • Barcelona • Belgrade • Novi Sad • Cluj • Manchester

Guest Editor
Bahil Ghanim
Department of General and
Thoracic Surgery
Karl Landsteiner University
of Health Sciences
Krems
Austria

Editorial Office
MDPI AG
Grosspeteranlage 5
4052 Basel, Switzerland

This is a reprint of the Special Issue, published open access by the journal *Cancers* (ISSN 2072-6694), freely accessible at: www.mdpi.com/journal/cancers/special_issues/Solitary_Fibrous_Tumor.

For citation purposes, cite each article independently as indicated on the article page online and using the guide below:

Lastname, A.A.; Lastname, B.B. Article Title. *Journal Name* **Year**, *Volume Number*, Page Range.

ISBN 978-3-7258-3700-7 (Hbk)
ISBN 978-3-7258-3699-4 (PDF)
https://doi.org/10.3390/books978-3-7258-3699-4

Contents

About the Editor . . . . **vii**

Bahil Ghanim
Solitary Fibrous Tumor
Reprinted from: *Cancers* **2024**, *16*, 3573, https://doi.org/10.3390/cancers16213573 . . . . **1**

Mattia Luca Piccinelli, Kyle Law, Reha-Baris Incesu, Stefano Tappero, Cristina Cano Garcia and Francesco Barletta et al.
Demographic and Clinical Characteristics of Malignant Solitary Fibrous Tumors: A SEER Database Analysis
Reprinted from: *Cancers* **2024**, *16*, 3331, https://doi.org/10.3390/cancers16193331 . . . . **3**

Pietro Bertoglio, Giulia Querzoli, Peter Kestenholz, Marco Scarci, Marilina La Porta and Piergiorgio Solli et al.
Surgery for Solitary Fibrous Tumors of the Pleura: A Review of the Available Evidence
Reprinted from: *Cancers* **2023**, *15*, 4166, https://doi.org/10.3390/cancers15164166 . . . . **16**

Yi Li, John T. Nguyen, Manasvini Ammanamanchi, Zikun Zhou, Elijah F. Harbut and Jose L. Mondaza-Hernandez et al.
Reduction of Tumor Growth with RNA-Targeting Treatment of the NAB2–STAT6 Fusion Transcript in Solitary Fibrous Tumor Models
Reprinted from: *Cancers* **2023**, *15*, 3127, https://doi.org/10.3390/cancers15123127 . . . . **33**

Mohammad Hassani, Sungmi Jung, Elaheh Ghodsi, Leila Seddigh, Paul Kooner and Ahmed Aoude et al.
Value of Cellular Components and Focal Dedifferentiation to Predict the Risk of Metastasis in a Benign-Appearing Extra-Meningeal Solitary Fibrous Tumor: An Original Series from a Tertiary Sarcoma Center
Reprinted from: *Cancers* **2023**, *15*, 1441, https://doi.org/10.3390/cancers15051441 . . . . **49**

Connor J. Kinslow, Ali I. Rae, Prashanth Kumar, Guy M. McKhann, Michael B. Sisti and Jeffrey N. Bruce et al.
Risk Stratification for Management of Solitary Fibrous Tumor/Hemangiopericytoma of the Central Nervous System
Reprinted from: *Cancers* **2023**, *15*, 876, https://doi.org/10.3390/cancers15030876 . . . . **59**

Marine Lottin, Alexandre Escande, Luc Bauchet, Marie Albert-Thananayagam, Maël Barthoulot and Matthieu Peyre et al.
Intracranial Solitary Fibrous Tumour Management: A French Multicentre Retrospective Study
Reprinted from: *Cancers* **2023**, *15*, 704, https://doi.org/10.3390/cancers15030704 . . . . **73**

Bahil Ghanim, Dina Baier, Christine Pirker, Leonhard Müllauer, Katharina Sinn and Gyoergy Lang et al.
Trabectedin Is Active against Two Novel, Patient-Derived Solitary Fibrous Pleural Tumor Cell Lines and Synergizes with Ponatinib
Reprinted from: *Cancers* **2022**, *14*, 5602, https://doi.org/10.3390/cancers14225602 . . . . **89**

Caroline Apra, Amira El Arbi, Anne-Sophie Montero, Fabrice Parker and Steven Knafo
Spinal Solitary Fibrous Tumors: An Original Multicenter Series and Systematic Review of Presentation, Management, and Prognosis
Reprinted from: *Cancers* **2022**, *14*, 2839, https://doi.org/10.3390/cancers14122839 . . . . **105**

Axel de Bernardi, Armelle Dufresne, Florence Mishellany, Jean-Yves Blay, Isabelle Ray-Coquard and Mehdi Brahmi
Novel Therapeutic Options for Solitary Fibrous Tumor: Antiangiogenic Therapy and Beyond
Reprinted from: *Cancers* **2022**, *14*, 1064, https://doi.org/10.3390/cancers14041064 . . . . **116**

Giuseppe Bianchi, Debora Lana, Marco Gambarotti, Cristina Ferrari, Marta Sbaraglia and Elena Pedrini et al.
Clinical, Histological, and Molecular Features of Solitary Fibrous Tumor of Bone: A Single Institution Retrospective Review
Reprinted from: *Cancers* **2021**, *13*, 2470, https://doi.org/10.3390/cancers13102470 . . . . **134**

About the Editor

Bahil Ghanim

The guest editor Bahil Ghanim is working as a thoracic surgeon and researcher at the Karl Landsteiner University of Health Sciences in Krems, Austria. His main research focus is on thoracic malignancies, rare diseases in thoracic ongology, and thoracic surgery.

MDPI

Editorial

Solitary Fibrous Tumor

Bahil Ghanim [1,2]

[1] Karl Landsteiner University of Health Sciences, Dr. Karl-Dorrek-Straße 30, 3500 Krems, Austria; bahil.ghanim@kl.ac.at

[2] Department of General and Thoracic Surgery, University Hospital Krems, Mitterweg 10, 3500 Krems, Austria

Solitary fibrous tumor (SFT) is an orphan disease of mesenchymal origin. The tumor can occur anywhere in the human body but is most frequently found in the chest. Surgery remains the standard of care in all SFTs. However, even completely resected disease can recur many years after surgery and its clinical behavior is unpredictable [1]. Genetic characterization using NAB2-STAT6 fusion has helped to better define and thus understand the disease for over a decade [2]. In addition, some risk stratification models have been developed and validated to estimate the clinical outcome [3].

Regarding therapy, surgery remains the cornerstone in SFT management but systemic treatment (including antiangiogenetic therapy) as well as radiotherapy have been suggested to improve survival and quality of life, especially in advanced stages of the disease [4]. Despite the current advantages of non-surgical SFT therapy, there is still an urgent need to enhance our clinical and biological understanding of this rare malignancy.

The present Special Issue aims to improve—for patients as well as for physicians —the frustrating situation in which an established and effective therapy and follow-up strategies are still lacking, and focuses on providing an overview of the current standard of care by contributing three review articles.

In addition, seven research articles enhance the knowledge regarding risk stratification and novel treatment approaches. On the one hand, a better risk stratification is urgently needed to establish follow-up strategies to treat this orphan disease in a clinically meaningful manner. On the other hand, an effective therapy that could provide an alternative to surgery is of interest for non-resectable patients, as well as in the setting of multimodal therapy before and/or after surgery. We hope that this Special Issue will help to improve the outcome of this rare and challenging disease.

Funding: This Editorial was not supported by funding.

Acknowledgments: As Guest Editor of the Special Issue "Solitary fibrous tumor", I want to thank all authors who spend their time and effort on exploring an orphan tumor to improve the situation of our patients and thus making this edition possible.

Conflicts of Interest: The author declares no conflicts of interest.

List of Contributions:

1. Bianchi, G.; Lana, D.; Gambarotti, M.; Ferrari, C.; Sbaraglia, M.; Pedrini, E.; Pazzaglia, L.; Sangiorgi, L.; Bartolotti, I.; Dei Tos, A.; et al. Clinical, Histological, and Molecular Features of Solitary Fibrous Tumor of Bone: A Single Institution Retrospective Review. *Cancers* **2021**, *13*, 2470. https://doi.org/10.3390/cancers13102470.

2. de Bernardi, A.; Dufresne, A.; Mishellany, F.; Blay, J.; Ray-Coquard, I.; Brahmi, M. Novel Therapeutic Options for Solitary Fibrous Tumor: Antiangiogenic Therapy and Beyond. *Cancers* **2022**, *14*, 1064. https://doi.org/10.3390/cancers14041064.

3. Apra, C.; El Arbi, A.; Montero, A.; Parker, F.; Knafo, S. Spinal Solitary Fibrous Tumors: An Original Multicenter Series and Systematic Review of Presentation, Management, and Prognosis. *Cancers* **2022**, *14*, 2839. https://doi.org/10.3390/cancers14122839.

check for updates

Citation: Ghanim, B. Solitary Fibrous Tumor. *Cancers* **2024**, *16*, 3573. https://doi.org/10.3390/cancers16213573

Received: 8 October 2024
Accepted: 15 October 2024
Published: 23 October 2024

4. Ghanim, B.; Baier, D.; Pirker, C.; Müllauer, L.; Sinn, K.; Lang, G.; Hoetzenecker, K.; Berger, W. Trabectedin Is Active against Two Novel, Patient-Derived Solitary Fibrous Pleural Tumor Cell Lines and Synergizes with Ponatinib. *Cancers* **2022**, *14*, 5602. https://doi.org/10.3390/cancers14225602.

5. Lottin, M.; Escande, A.; Bauchet, L.; Albert-Thananayagam, M.; Barthoulot, M.; Peyre, M.; Boone, M.; Zouaoui, S.; Guyotat, J.; Penchet, G.; et al. Intracranial Solitary Fibrous Tumour Management: A French Multicentre Retrospective Study. *Cancers* **2023**, *15*, 704. https://doi.org/10.3390/cancers15030704.

6. Kinslow, C.; Rae, A.; Kumar, P.; McKhann, G.; Sisti, M.; Bruce, J.; Yu, J.; Cheng, S.; Wang, T. Risk Stratification for Management of Solitary Fibrous Tumor/Hemangiopericytoma of the Central Nervous System. *Cancers* **2023**, *15*, 876. https://doi.org/10.3390/cancers15030876.

7. Hassani, M.; Jung, S.; Ghodsi, E.; Seddigh, L.; Kooner, P.; Aoude, A.; Turcotte, R. Value of Cellular Components and Focal Dedifferentiation to Predict the Risk of Metastasis in a Benign-Appearing Extra-Meningeal Solitary Fibrous Tumor: An Original Series from a Tertiary Sarcoma Center. *Cancers* **2023**, *15*, 1441. https://doi.org/10.3390/cancers15051441.

8. Li, Y.; Nguyen, J.; Ammanamanchi, M.; Zhou, Z.; Harbut, E.; Mondaza-Hernandez, J.; Meyer, C.; Moura, D.; Martin-Broto, J.; Hayenga, H.; et al. Reduction of Tumor Growth with RNA-Targeting Treatment of the NAB2–STAT6 Fusion Transcript in Solitary Fibrous Tumor Models. *Cancers* **2023**, *15*, 3127. https://doi.org/10.3390/cancers15123127.

9. Bertoglio, P.; Querzoli, G.; Kestenholz, P.; Scarci, M.; La Porta, M.; Solli, P.; Minervini, F. Surgery for Solitary Fibrous Tumors of the Pleura: A Review of the Available Evidence. *Cancers* **2023**, *15*, 4166. https://doi.org/10.3390/cancers15164166.

10. Piccinelli, M.; Law, K.; Incesu, R.; Tappero, S.; Cano Garcia, C.; Barletta, F.; Morra, S.; Scheipner, L.; Baudo, A.; Tian, Z.; et al. Demographic and Clinical Characteristics of Malignant Solitary Fibrous Tumors: A SEER Database Analysis. *Cancers* **2024**, *16*, 3331. https://doi.org/10.3390/cancers16193331.

References

1. Thway, K.; Ng, W.; Noujaim, J.; Jones, R.L.; Fisher, C. The Current Status of Solitary Fibrous Tumor: Diagnostic Features, Variants, and Genetics. *Int. J. Surg. Pathol.* **2016**, *24*, 281–292. [CrossRef] [PubMed]

2. Robinson, D.R.; Wu, Y.M.; Kalyana-Sundaram, S.; Cao, X.; Lonigro, R.J.; Sung, Y.S.; Chen, C.L.; Zhang, L.; Wang, R.; Su, F.; et al. Identification of recurrent NAB2-STAT6 gene fusions in solitary fibrous tumor by integrative sequencing. *Nat. Genet.* **2013**, *45*, 180–185. [CrossRef] [PubMed]

3. Demicco, E.G.; Park, M.S.; Araujo, D.M.; Fox, P.S.; Bassett, R.L.; Pollock, R.E.; Lazar, A.J.; Wang, W.L. Solitary fibrous tumor: A clinicopathological study of 110 cases and proposed risk assessment model. *Mod. Pathol.* **2012**, *25*, 1298–1306. [CrossRef] [PubMed]

4. Ren, C.; D'Amato, G.; Hornicek, F.J.; Tao, H.; Duan, Z. Advances in the Molecular Biology of the Solitary Fibrous Tumor and Potential Impact on Clinical Applications. *Cancer Metastasis Rev.* 2024; *Online ahead of print.* [CrossRef]

Article

Demographic and Clinical Characteristics of Malignant Solitary Fibrous Tumors: A SEER Database Analysis

Mattia Luca Piccinelli [1,2,3,]*, Kyle Law [1], Reha-Baris Incesu [1,4], Stefano Tappero [1,5,6], Cristina Cano Garcia [1,7],
Francesco Barletta [1,8], Simone Morra [1,9], Lukas Scheipner [1,10], Andrea Baudo [1,3,11], Zhe Tian [1],
Stefano Luzzago [2,12], Francesco Alessandro Mistretta [2,12], Matteo Ferro [2], Fred Saad [1],
Shahrokh F. Shariat [13,14,15,16], Luca Carmignani [11,17], Sascha Ahyai [10], Nicola Longo [9], Alberto Briganti [8],
Felix K. H. Chun [7], Carlo Terrone [5,6], Derya Tilki [4,18,19], Ottavio de Cobelli [2,12], Gennaro Musi [2,12]
and Pierre I. Karakiewicz [1]

1. Cancer Prognostics and Health Outcomes Unit, Division of Urology, University of Montréal Health Center, Montréal, QC H2X 3E4, Canada
2. Department of Urology, IEO European Institute of Oncology, IRCCS, Via Ripamonti 435, 20141 Milan, Italy
3. School of Urology, Università degli Studi di Milano, 20122 Milan, Italy
4. Martini-Klinik Prostate Cancer Center, University Hospital Hamburg-Eppendorf, 20251 Hamburg, Germany
5. Department of Urology, IRCCS Policlinico San Martino, 16132 Genova, Italy
6. Department of Surgical and Diagnostic Integrated Sciences (DISC), University of Genova, 16126 Genova, Italy
7. Department of Urology, University Hospital Frankfurt, Goethe University Frankfurt am Main, 60629 Frankfurt am Main, Germany
8. Division of Experimental Oncology/Unit of Urology, URI, Urological Research Institute, IRCCS San Raffaele Scientific Institute, 20132 Milan, Italy
9. Department of Neurosciences, Science of Reproduction and Odontostomatology, University of Naples Federico II, 80131 Naples, Italy
10. Department of Urology, Medical University of Graz, 8010 Graz, Austria
11. Department of Urology, IRCCS Policlinico San Donato, 20097 Milan, Italy
12. Department of Oncology and Haemato-Oncology, Università degli Studi di Milano, 20122 Milan, Italy
13. Department of Urology, Comprehensive Cancer Center, Medical University of Vienna, 1090 Vienna, Austria
14. Department of Urology, Weill Cornell Medical College, New York, NY 10021, USA
15. Department of Urology, University of Texas Southwestern Medical Center, Dallas, TX 75390, USA
16. Hourani Center of Applied Scientific Research, Al-Ahliyya Amman University, Amman P.O. Box 19328, Jordan
17. Department of Urology, Istituto di Ricovero e Cura a Carattere Scientifico Ospedale Galeazzi—Sant'Ambrogio, 20157 Milan, Italy
18. Department of Urology, University Hospital Hamburg-Eppendorf, 20251 Hamburg, Germany
19. Department of Urology, Koc University Hospital, 20251 Istanbul, Turkey
* Correspondence: ieo5209@ieo.it

Citation: Piccinelli, M.L.; Law, K.; Incesu, R.-B.; Tappero, S.; Cano Garcia, C.; Barletta, F.; Morra, S.; Scheipner, L.; Baudo, A.; Tian, Z.; et al. Demographic and Clinical Characteristics of Malignant Solitary Fibrous Tumors: A SEER Database Analysis. *Cancers* 2024, *16*, 3331. https://doi.org/10.3390/cancers 16193331

Academic Editor: Bahil Ghanim

Received: 29 July 2024
Revised: 22 September 2024
Accepted: 27 September 2024
Published: 29 September 2024

Simple Summary: Solitary fibrous tumors represent a rare mesenchymal malignancy that can occur anywhere in the body. Due to the low prevalence of the disease, there is a lack of contemporary data regarding patient demographics and cancer-control outcomes. We validated the importance of stage and surgical resection as independent predictors of cancer-specific mortality in malignant solitary fibrous tumors. Moreover, we provide novel observations regarding the independent importance of tumor size, regardless of the site of origin, stage and/or surgical resection status.

Abstract: Background/Objectives: Solitary fibrous tumors (SFTs) represent a rare mesenchymal malignancy that can occur anywhere in the body. Due to the low prevalence of the disease, there is a lack of contemporary data regarding patient demographics and cancer-control outcomes. Methods: Within the SEER database (2000–2019), we identified 1134 patients diagnosed with malignant SFTs. The distributions of patient demographics and tumor characteristics were tabulated. Cumulative incidence plots and competing risks analyses were used to estimate cancer-specific mortality (CSM) after adjustment for other-cause mortality. Results: Of 1134 SFT patients, 87% underwent surgical resection. Most of the tumors were in the chest (28%), central nervous system (22%), head and neck (11%), pelvis (11%), extremities (10%), abdomen (10%) and retroperitoneum (6%), in that order. Stage was distributed as follows: localized (42%) vs. locally advanced (35%) vs. metastatic (13%). In multivariable competing risks models, independent predictors of higher CSM were stage (locally advanced HR: 1.6; metastatic HR: 2.9), non-surgical management (HR: 3.6) and tumor size (9–15.9 cm HR: 1.6; ≥16 cm HR: 1.9). Conclusions: We validated the importance of stage and surgical resection as independent predictors of CSM in malignant SFTs. Moreover, we provide novel observations regarding the independent importance of tumor size, regardless of the site of origin, stage and/or surgical resection status.

Keywords: solitary fibrous tumor; competing risks analyses; tumor-size cut-offs

1. Introduction

Solitary fibrous tumors (SFTs) represent a rare mesenchymal malignancy that accounts for <2% of all soft-tissue sarcomas and can occur anywhere in the body [1–3]. Cellular tumors, which were formerly known as hemangiopericytomas, are now considered to be part of the SFT spectrum [2,4–8]. Although indicators of more aggressive treated natural history consist of elevated mitotic index, infiltrative margins, hypercellularity, pleomorphism and necrosis (specifically, a proposed definition of malignant SFTs is tumors with focal areas of marked increased cellularity described as greater than 5% of tumor that are devoid of alternating sclerotic areas and have greater than four mitoses per ten high-powered fields [1]), no consensus exists regarding the treated natural history when such features are absent from pre-treatment biopsy specimens. In consequence, the search for accurate and reliable predictors of treated natural history in SFTs is ongoing [2–4,9,10]. In that regard, only small case series (n = 110–219) have been published, and these suggest a 4–19% rate of 10-year local recurrence and a 13–45% rate of 10-year metastatic progression [6,11,12]. Similarly, survival patterns according to the site of origin, stage and surgical resection status are also based on very limited data [6,12–17]. Last but not least, no systematic assessment of tumor-size cut-offs for the prediction of cancer-specific mortality (CSM) has ever been performed to date. We addressed these knowledge gaps, relying on the 2000–2019 Surveillance, Epidemiology, and End Results (SEER) database [18]. We tested whether stage, surgical resection and possibly tumor size represent predictors of CSM across different sites of origin of primary malignant SFTs. Moreover, we hypothesized that significant differences in patient characteristics and CSM rates exist according to the site of origin, stage, surgical resection status and tumor size.

2. Materials and Methods

2.1. Patient Characteristics

Within the SEER database (2000–2019), we focused on patients aged ≥18 who harbored malignant SFTs (International Classification of Disease for Oncology histology code 8815/3 and 9150/3 [4,7]) and had known follow-up and primary site. SFT origin was tabulated according to SEER location (central nervous system, extremities, head and neck, chest, pelvis, abdomen and retroperitoneum [6,11–14,17,19]) and SEER stage (localized, locally advanced and metastatic [14,15,17,18,20]). Specifically, SEER staging defines localized cancer as that limited to the organ in which it began, without evidence of spread. SEER staging defines locally advanced (or regional) cancer as that which has spread beyond the primary site to nearby lymph nodes or organs and tissues. Metastatic (or distant) cancer is defined as a disease that has spread from the primary site to distant organs or distant lymph nodes. Tumor size was stratified as follows: <9 cm, 9–15.9 cm and ≥16 cm.

2.2. Statistical Analysis

Descriptive statistics were used to characterize patient age and tumor size and stage. Cumulative incidence plots depicted CSM rates after adjustment for other-cause mortality (OCM). Moreover, we tested for the ideal tumor-size cut-off for the prediction of CSM using a minimum p-value approach. Subsequently, univariable and multivariable competing risks regression models were used to test for independent predictors of CSM after adjustment for OCM. All statistical tests were two-sided, with the level of significance set at $p < 0.05$, and were performed with R Software Environment for Statistical Computing and Graphics (R version 4.1.3, R Foundation for Statical Computing, Vienna Austria) [21].

3. Results

3.1. Patient and Tumor Characteristics in the Overall Cohort

Of 1134 patients with malignant SFTs, 551 (49%) were male and 989 (87%) were surgically treated (Table 1).

Table 1. Descriptive characteristics of patients diagnosed with malignant solitary fibrous tumors between 2000 and 2019, as recorded in the Surveillance, Epidemiology, and End Results database. Data are shown as medians for continuous variables or as counts and percentages (%) for categorical variables. IQR: interquartile range.

Malignant Solitary Fibrous Tumor	Overall $n = 1134$
Age at diagnosis (years) Median (IQR)	60 (50–69)
Sex—Male	551 (49%)
Race/ethnicity	
Caucasian	771 (68%)
African American	95 (8%)
Hispanic	149 (13%)
Asian/Pacific Islander	102 (9%)
Other	17 (2%)
Surgical resection	989 (87%)
Site of origin	
Extremities and head	500 (44%)
Central nervous system	261 (23%)
Head and neck	120 (11%)
Extremities	119 (10%)
Chest	322 (29%)
Infradiaphragmatic	312 (28%)
Pelvis	128 (11%)
Abdomen	114 (10%)
Retroperitoneum	70 (6%)
Size (cm)	
Median (IQR)	75 (46–120)
<9 cm	518 (46%)
9–15.9 cm	247 (22%)
≥16 cm	100 (9%)
Unknown	269 (24%)
Stage	
Localized	475 (42%)
Locally advanced	401 (35%)
Metastatic	142 (13%)
Unstaged	116 (10%)

The median age at diagnosis was 60 years (Figure 1a).

Most SFTs were located in the chest ($n = 322$, 29%), central nervous system ($n = 261$, 23%), head and neck ($n = 120$, 11%), pelvis ($n = 128$, 11%), extremities ($n = 119$, 10%), abdomen ($n = 114$, 10%) and retroperitoneum ($n = 70$, 6%), in that order (Supplementary Table S1). Overall, 475 (42%) of patients harbored localized tumors, while 401 (35%) had tumors that were locally advanced and 142 (13%) had tumors in the metastatic stage. In 116 (10%) patients, the stage was unknown. Median tumor size was 75 mm (IQR: 46–120).

a

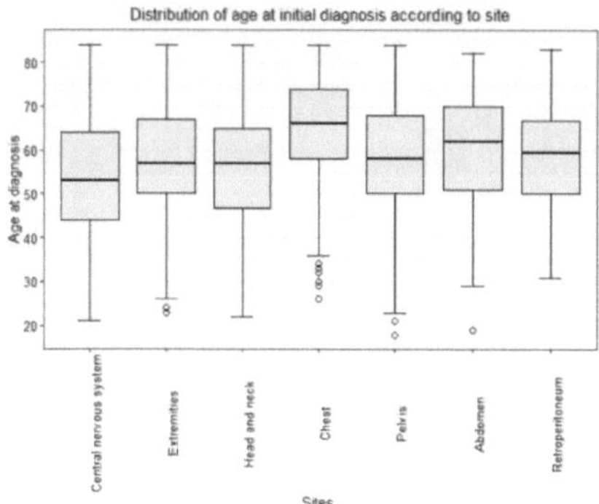

Site	Age at initial diagnosis	p-value
Overall	60 (50-69)	
Central nervous system	53 (44-64)	
Extremities	57 (50-67)	
Head and neck	57 (47-65)	
Chest	66 (58-74)	<0.001
Pelvis	58 (50-68)	
Abdomen	62 (51-70)	
Retroperitoneum	60 (50-67)	

b

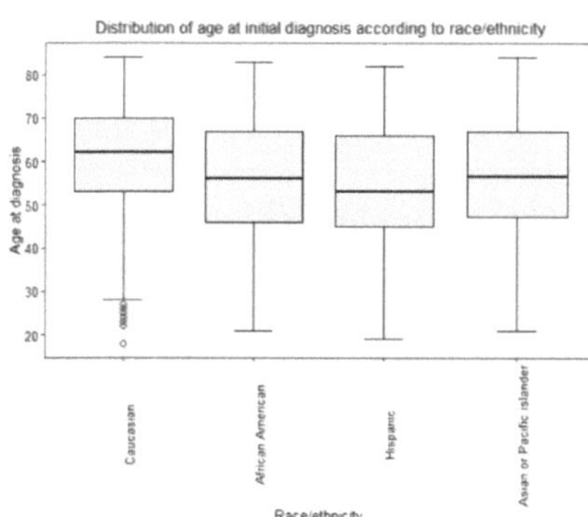

Site	Age at initial diagnosis	p-value
Overall	60 (50-69)	
Caucasian	62 (53-70)	
African American	56 (46-67)	<0.001
Hispanic	53 (45-66)	
Asian or Pacific islander	56 (47-67)	

c

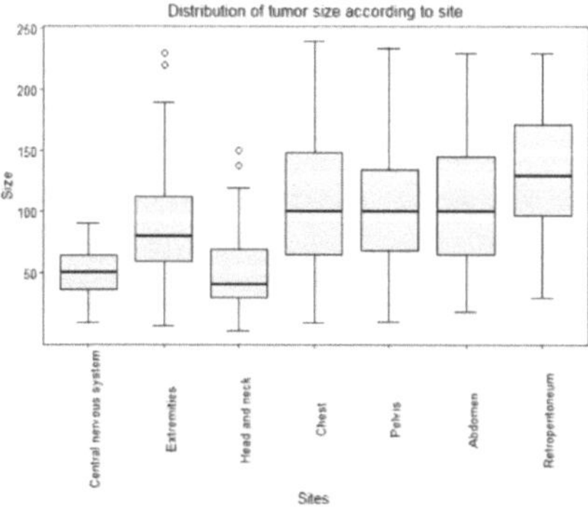

Site	Size at initial diagnosis (mm)	p-value
Overall	75 (46-120)	
Central nervous system	50 (36-64)	
Extremities	80 (59-112)	
Head and neck	40 (30-69)	
Chest	100 (65-149)	<0.001
Pelvis	100 (69-135)	
Abdomen	100 (65-145)	
Retroperitoneum	130 (97-172)	

Figure 1. Whisker plots depicting the distribution of patients diagnosed with malignant solitary fibrous tumors, as recorded in the 2000–2019 Surveillance, Epidemiology, and End Results database: (**a**) age at diagnosis according to tumor site of origin; (**b**) age at diagnosis according to race/ethnicity; (**c**) tumor size (mm) according to tumor site of origin.

3.2. Patient and Tumor Characteristics According to the Site of Origin

Differences in stage distribution were recorded according to the site of origin. SFTs in the extremities were the most frequently localized (*n* = 77, 72%, Figure 2a).

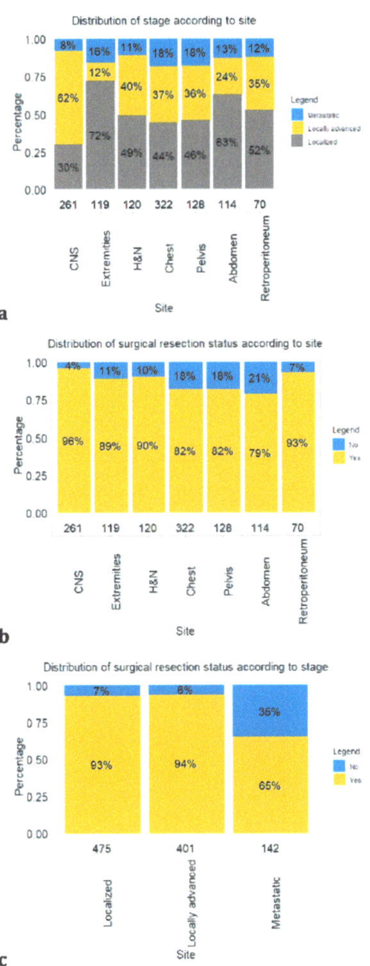

Figure 2. Bar plot depicting the distribution of patients diagnosed with malignant solitary fibrous tumors, as recorded in the 2000–2019 Surveillance, Epidemiology, and End Results database: (**a**) stage according to site; (**b**) surgical resection according to site; (**c**) surgical resection according to stage.

Conversely, SFTs in the central nervous system were the most frequently locally advanced (n = 152, 62%). The frequency of metastatic SFTs ranged from 8 (n = 20, central nervous system) to 18% (n = 51 in the chest and n = 21 in the pelvis). Rates of surgical resection status also varied according to the site of origin (Figure 2b). Specifically, 251 (96%) SFTs were surgically resected from the central nervous system, which was followed in frequency of resection by the retroperitoneum (n = 65, 93%), head and neck (n = 108, 90%), extremities (n = 106, 89%), chest (n = 264, 82%), pelvis (n = 105, 82%) and abdomen (n = 90, 79%), in that order. Finally, according to stage, 444 (93%) patients with localized SFTs vs. 375 (94%) patients with locally advanced SFTs vs. 92 (65%) patients with metastatic SFTs underwent surgical resection (Figure 2c). In general, the smaller SFTs had the head and neck (median size: 40 mm) and central nervous system (median size: 50 mm) as sites of origin. Conversely, the larger tumors had the chest (median size: 100 mm), pelvis (median size: 100 mm) and abdomen (median size: 100 mm) as sites of origin. Finally, the largest SFTs originated in the retroperitoneum (median size: 130 mm, Figure 1c).

3.3. Cancer-Specific and Other-Cause Mortality in Solitary Fibrous Tumor

In cumulative incidence plots, based on 1134 malignant SFT patients, 10-year CSM and OCM rates were 34 and 18%, respectively (Figure 3a).

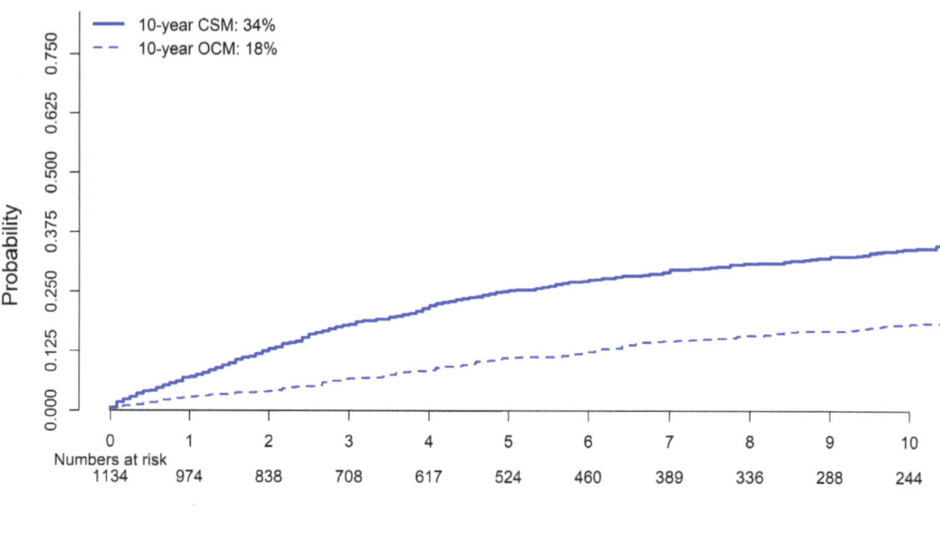

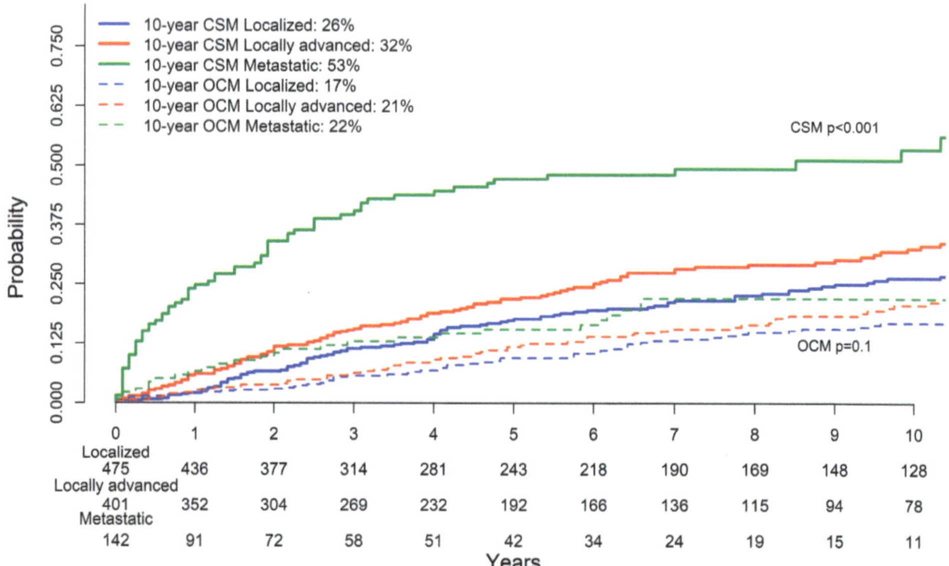

Figure 3. *Cont.*

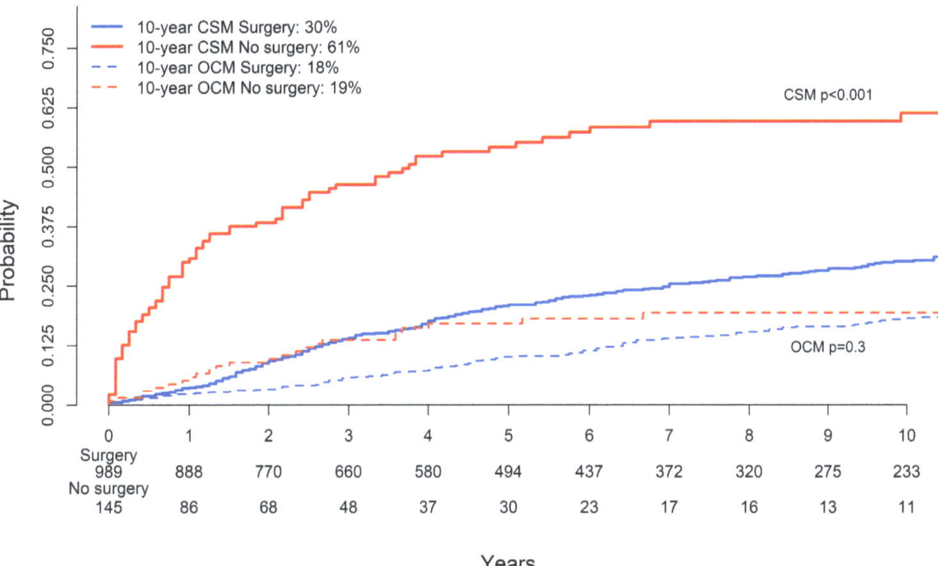

c

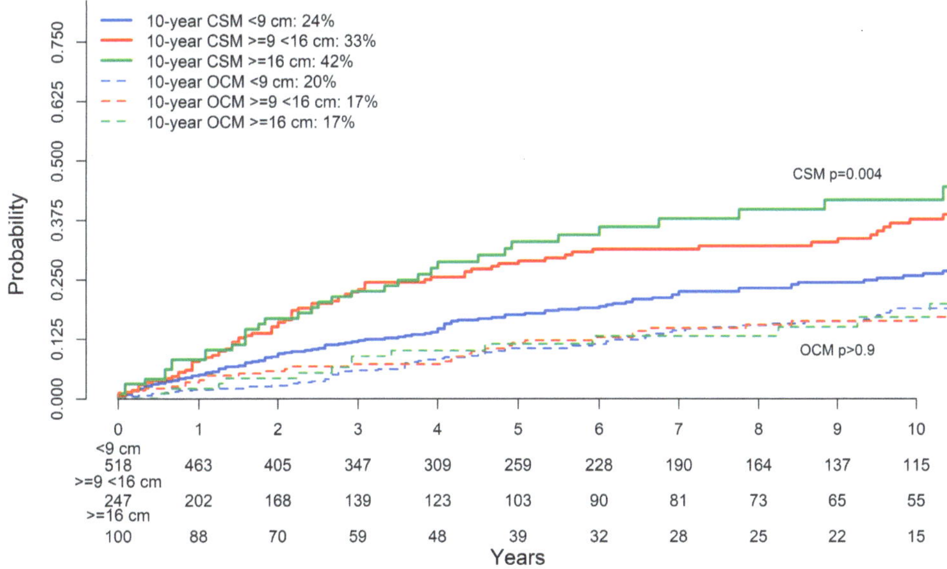

d

Figure 3. Cumulative incidence plots depicting cancer-specific mortality and other-cause mortality over 10 years in patients diagnosed with malignant solitary fibrous tumors in 2000–2019 according to the Surveillance, Epidemiology, and End Results database (**a**) overall and according to (**b**) stage; (**c**) surgical resection status; (**d**) tumor size.

Specifically, the lowest 10-year CSM rate was recorded for SFTs located in the head and neck (21%), with the second-lowest rate being that for SFTs located in the central nervous system (26%). Conversely, the highest rate of 10-year CSM was recorded for SFTs located in the abdomen (44%, Supplementary Figure S1a). CSM rates at ten years according to stage were 26% in localized vs. 32% in locally advanced vs. 53% in metastatic SFTs (Figure 3b). Ten-year CSM rates according to surgical resection status (yes vs. no) were 30 vs. 61%, respectively (Figure 3c). Finally, 10-year CSM rates according to tumor-size intervals were as follows: tumor size <9 cm 24%; tumor size 9–15.9 33%; tumor size $\geq$16 cm 42% (Figure 3d). In multivariable competing risks analyses, locally advanced stage (hazard ratio [HR]: 1.6, $p < 0.001$), metastatic stage (HR: 2.9, $p < 0.001$), non-surgical management (HR: 3.6, $p < 0.001$) and tumor size (9–15.9 cm HR: 1.6, $p = 0.01$; $\geq$16 cm HR: 1.9, $p = 0.01$) independently predicted higher CSM rates after additional adjustment for OCM (Table 2).

Table 2. Univariable and multivariable competing risks analyses predicting cancer-specific mortality and accounting for other-cause mortality. All patients were diagnosed with malignant solitary fibrous tumors between 2000 and 2019, as recorded in the Surveillance, Epidemiology, and End Results database.

Variables Tested	Univariable			Multivariable		
	Hazard Ratio	95% CI	*p*-Value	Hazard Ratio	95% CI	*p*-Value
Age at diagnosis (years)	1.01	(1.01–1.02)	<0.001	1.01	(1–1.02)	0.08
Sex—Female	0.92	(0.7–1.2)	0.5			
Race/ethnicity						
Caucasian	Ref					
African American	1.2	(0.7–1.9)	0.5			
Hispanic	0.7	(0.4–1.1)	0.1			
Asian or Pacific Islander	0.9	(0.6–1.4)	0.7			
Surgical resection status—No	4	(2.7–6.1)	<0.001	3.6	(2.3–5.6)	<0.001
Site of origin						
Central nervous system	Ref			Ref		
Extremities	1.6	(0.98–2.7)	0.06	1.6	(0.9–2.9)	0.1
Head and neck	1.3	(0.7–2.2)	0.4	1.2	(0.7–2.1)	0.6
Chest	1.7	(1.1–2.5)	0.01	0.97	(0.6–1.7)	0.9
Pelvis	1.5	(0.9–2.5)	0.11	0.9	(0.5–1.7)	0.8
Abdomen	1.7	(0.99–3)	0.06	1.3	(0.7–2.5)	0.5
Retroperitoneum	1.3	(0.6–2.5)	0.52	0.8	(0.3–1.6)	0.5
Size						
<9 cm	Ref			Ref		
9–15.9 cm	1.5	(1.1–2.0)	0.01	1.6	(1.1–2.4)	0.01
$\geq$16 cm	1.8	(1.2–2.6)	<0.001	1.9	(1.1–3.1)	0.01
Stage						
Localized	Ref			Ref		
Locally advanced	1.5	(1.1–2)	0.02	1.6	(1.2–2.3)	<0.001
Metastatic	3.4	(2.3–5)	<0.001	2.9	(2.0–4.4)	<0.001

Conversely, after adjustment for age at diagnosis, stage, surgical resection status and tumor size and additional adjustment for OCM, the site of origin failed to achieve independent predictor status for CSM. Finally, in separate multivariable competing risks analyses testing surgical resection status according to stage, non-surgical management achieved independent predictor status for higher CSM in localized (HR: 1.8, $p = 0.03$), locally advanced (HR: 2.6, $p = 0.01$) and metastatic (HR: 5.1, $p < 0.001$) SFTs (Table 3).

Table 3. Separate multivariable competing risks analyses testing the independent CSM predictor status of surgical resection status after adjustment for OCM according to stage. All models were adjusted for age at diagnosis, site of origin and tumor size. All patients were diagnosed with malignant solitary fibrous tumors between 2000 and 2019, as recorded in the Surveillance, Epidemiology, and End Results database.

	Localized			Locally Advanced			Metastatic		
Variables Tested	Hazard Ratio	95% CI	*p*-Value	Hazard Ratio	95% CI	*p*-Value	Hazard Ratio	95% CI	*p*-Value
Surgical resection status—No	1.8	(0.6–5.1)	0.03	2.6	(1.3–5.3)	0.01	5.1	(2.6–9.8)	<0.001

4. Discussion

Small series studies have explored SFT survival patterns according to the site of origin, stage, surgical resection status and tumor size [6,12–17]. However, no comprehensive and systematic assessment of tumor-size cut-offs for the prediction of CSM has ever been reported. We addressed these knowledge gaps and hypothesized that significant differences in patient characteristics and CSM rates exist according to the site of origin, stage, surgical resection status and tumor size. Several important observations were made.

First, we provided the most detailed tabulation of sites of origin within the largest ($n = 1134$) and most contemporary (year of diagnosis: 2000–2019) cohort of patients with malignant SFTs. We identified the chest (28%) as the most frequent site of origin, followed by the central nervous system (22%). The rates of SFT origin were virtually equally distributed between head and neck (11%), pelvis (11%), extremities (10%) and abdomen (10%). Conversely, retroperitoneal SFTs were the least frequent (6%). Based on the absence of previously published detailed data regarding sites of SFT origin, our observations can only be partially compared to other smaller and more historical series [6,11–14,16,17,19,22–24]. However, these comparisons revealed a close agreement between the current rates and historical rates from smaller series.

The current database also allowed us to tabulate SFTs according to the stage and provided the most robust and contemporary results. Specifically, of all SFTs, 42% were localized, 35% were locally advanced and 13% were metastatic. Unfortunately, we also observed that 10% were of unknown stage. The percentage of SFTs of unknown stage is comparable to percentages of tumors of unknown stage in the SEER database for other malignancies such as kidney [25] or prostate [26] cancers. The distribution of SFT patients across stages differed from that found in the study of Wang et al. [15] ($n = 1243$, year of diagnosis: 1975–2016), which relied on a more historical SEER cohort. For example, in the analyses by Wang et al., only 17% of patients harbored tumors in the locally advanced stage vs. 35% in the current cohort. This discrepancy may be explained by a very elevated rate of tumors of unknown stage in the Wang et al. cohort: 35%, vs. 10% in the current study.

Finally, we provided the most generalizable distribution of surgical resection status. Overall, 87% of patients had undergone surgery. This observation is in perfect agreement with that in the historical cohort of Wang et al. [15], where surgical resection was accomplished in 88% of SFT patients. However, in the Wang et al. study, surgical resection status was not stratified according to the SFT site of origin and stage. We addressed this knowledge gap in the current study. Specifically, the highest rate of surgical resection was recorded for tumors in the central nervous system (96%), followed by those in the retroperitoneum (93%), head and neck (90%), extremities (89%), chest (82%), pelvis (82%) and abdomen (79%), in that order. We also provided rates of surgical resection status according to stage. Specifically, 444/475 (93%) patients with localized SFTs underwent surgical resection vs. 375/401 (94%) patients with locally advanced SFTs and 92/142 (65%) patients with metastatic SFTs. The very elevated surgical resection rates associated with localized and locally advanced SFTs (93–94%) validate the pivotal role of surgery. Additionally, the central role of surgery was also confirmed in patients with metastatic SFTs, the vast majority (65%) of whom underwent resection.

Second, we are the first to validate that the absence of surgical resection independently predicts higher CSM (HR: 3.6, $p < 0.001$) in malignant SFTs of all stages after adjustment for OCM. Moreover, in stage-specific analyses, absence of surgical resection also independently predicted higher CSM. Specifically, for localized SFTs, absence of surgical resection exhibited an HR of 1.8 ($p = 0.03$); this value was 2.6 ($p = 0.01$) in the locally advanced stage and 5.1 ($p < 0.001$) in the metastatic stage. These observations validate the central role of surgical resection in the contemporary management of SFTs at all stages. Moreover, the increase in the magnitude of HRs from the localized to the locally advanced to the metastatic stages adds further evidence supporting the disadvantage of non-surgical management, which is most pronounced in metastatic patients.

Third, we also provided the most contemporary and generalizable validation of the importance of stage as a predictor of CSM. Here, relative to the localized stage, patients with locally advanced SFTs exhibited an HR of 1.6 ($p < 0.001$); this value was 2.9 ($p < 0.001$) in the metastatic stage. Unfortunately, direct comparisons with other series regarding the effect of the stage, as well as of the effect of surgical resection status according to specific SFT stage, cannot be made. For example, Wushou et al. [14] addressed only hemangiopericytoma, which represents only a subset of currently diagnosed malignant SFTs. Moreover, no previous studies relied on competing risks analyses adjusting for OCM when the stage was tested in a multivariable fashion.

Fourth, we are the first to perform a comprehensive and systematic assessment of tumor-size cut-offs for the prediction of CSM in malignant SFTs. Here, tumor-size cut-offs of <9, 9–15.9 and ≥16 cm emerged as ideal. Moreover, their independent predictor status was confirmed in multivariable analyses that included age at diagnosis, site of origin, stage, surgical resection status and additional adjustment for OCM. Our findings should ideally be validated within an equally large or even larger population-based data repository. Previous analyses regarding tumor size relied on the median [11,12,16,27] or generally accepted tumor-size cut-offs [28–30] used for other primary tumors, such as retroperitoneal sarcoma [31]. However, none of these previous smaller-scale ($n = 110$–239) analyses questioned the internal validity of such definitions for SFT patients. The most widely used and established cut-offs were defined by Demicco et al. [6] and are included in their scoring system to predict distant metastasis. Specifically, this innovative and elegant study provided the basis for the contemporary management of SFTs. The authors (Demicco et al.) relied on 110 SFT patients treated at M.D. Anderson Cancer Center (1986–2009). For analysis purposes, tumor-size cut-offs of <5, 5–9.9, 10–14.9 and ≥15 cm were used. However, these cut-off values were not based on specific clinical or statistical criteria. Instead, they may have been adopted from values used for other primary tumors such as retroperitoneal sarcoma [31]. Finally, the independent predictor status for these tumor-size cut-offs (<5, 5–9.9, 10–14.9 and ≥15 cm) was not tested regarding CSM. Similar methodological limitations regarding testing of tumor-size cut-offs also apply to the study by Gholami et al. [12]. Here, the authors relied on a single tumor-size cut-off of 8 cm within a historical cohort (1982–2015) of 219 SFT patients treated at Memorial Sloan Kettering Cancer Center. Importantly, the majority (74%) of the patients in the Gholami et al. study harbored non-malignant SFTs. In consequence, the proposed tumor-size cut-off is predominantly applicable to non-malignant SFTs, not to malignant SFTs.

Fifth, we tested for OCM rates and, although malignant SFT is associated with high rates of 10-year CSM, some patients died of other causes. Based on the absence of data quantifying OCM rates in SFTs, we provide values numbers as follows: 5-year OCM 11% and 10-year OCM 18%. Since OCM affects a non-negligible portion of SFT patients, ideally, competing risks analyses should be preferred when CSM rates are addressed.

Overall, the present study is based on the largest and most contemporary malignant SFT cohort analyzed to date and provides the most robust, comprehensive and detailed analyses of patient- and tumor-associated risk factors that may affect CSM. Three variables emerged as independent predictors of CSM in multivariable competing risks models that also adjusted for OCM: stage, surgical resection status and tumor size. We relied on a minimum p-value approach to explore potential tumor-size cut-offs. Ideal tumor-size cut-offs of <9, 9–15.9 and $\geq$15 emerged and represented independent predictors of CSM. However, these tumor-size cut-offs ideally require independent validation within an external cohort.

Despite its novelty, our study is not devoid of limitations. First, the SEER is a retrospective database with the potential for selection biases. However, observational databases such as SEER or NCDB represent the only opportunity to study rare primary tumors and reach statistically robust conclusions. Second, rates of local recurrence, metastatic progression and preoperative and postoperative treatments, as well as predictors of cancer-control outcomes (such as mitotic count, tumor necrosis and positive surgical margins) are not available in the SEER database. In consequence, our results should be tested and validated in other large-scale cohorts of patients with malignant SFTs. Fourth, SEER lacks specific baseline comorbidity information. In consequence, more detailed analyses adjusting for comorbidities were not possible. However, we partially addressed this limitation by the inclusion of OCM rates in our analyses.

5. Conclusions

We validated the importance of stage and surgical resection as independent predictors of cancer-specific mortality in malignant solitary fibrous tumors. Moreover, we provided novel observations regarding the independent importance of tumor size, regardless of the site of origin, stage and/or surgical resection status.

Supplementary Materials: The following supporting information can be downloaded at: https://www.mdpi.com/article/10.3390/cancers16193331/s1. Figure S1: Cumulative incidence plots and summary table depicting cancer-specific mortality and other-cause mortality over 10 years in patients with malignant solitary fibrous tumors recorded in the 2000–2019 Surveillance, Epidemiology, and End Results database according to (a) site, (b) location. Table S1: Distribution of organ of origin among 198 patients diagnosed with pelvis and retroperitoneum solitary fibrous tumor between 2000 and 2019, as recorded in the Surveillance, Epidemiology, and End Results database.

Author Contributions: M.L.P.: study concept, study design, data collection, data analysis, interpretation, writing the paper; K.L.: interpretation, writing the paper; R.-B.I.: data analysis, interpretation; S.T.: data analysis, interpretation; C.C.G.: data analysis, interpretation; F.B.: data analysis, interpretation; S.M.: data analysis, interpretation; L.S.: data analysis, interpretation; A.B. (Andrea Baudo): data analysis, interpretation; Z.T.: data analysis; S.L.: interpretation; F.A.M.: interpretation; M.F.: interpretation; F.S.: interpretation; S.F.S.: interpretation; L.C.: interpretation; S.A.: interpretation; N.L.: interpretation; A.B. (Alberto Briganti): interpretation; F.K.H.C.: interpretation; C.T.: interpretation; D.T.: interpretation; O.d.C.: interpretation; G.M.: interpretation; P.I.K.: study concept, study design, interpretation, writing the paper. All authors have read and agreed to the published version of the manuscript.

Funding: This research received no external funding.

Institutional Review Board Statement: Not applicable.

Informed Consent Statement: Not applicable.

Data Availability Statement: Data available in a publicly accessible repository.

Conflicts of Interest: The authors declare no conflicts of interest.

References

1. Gold, J.S.; Antonescu, C.R.; Hajdu, C.; Ferrone, C.R.; Hussain, M.; Lewis, J.J.; Brennan, M.F.; Coit, D.G. Clinicopathologic Correlates of Solitary Fibrous Tumors. *Cancer* **2002**, *94*, 1057–1068. [CrossRef] [PubMed]
2. de Bernardi, A.; Dufresne, A.; Mishellany, F.; Blay, J.Y.; Ray-Coquard, I.; Brahmi, M. Novel Therapeutic Options for Solitary Fibrous Tumor: Antiangiogenic Therapy and Beyond. *Cancers* **2022**, *14*, 1064. [CrossRef] [PubMed]

3. Machado, I.; Giner, F.; Cruz, J.; Lavernia, J.; Marhuenda-Fluixa, A.; Claramunt, R.; López-Guerrero, J.A.; Navarro, S.; Ferrandez, A.; Bujeda, Á.B.; et al. Extra-Meningeal Solitary Fibrous Tumor: An Evolving Entity with Chameleonic Morphological Diversity, a Hallmark Molecular Alteration and Unresolved Issues in Risk Stratification Assessment. *Histol. Histopathol.* **2023**, *38*, 1079–1097. [CrossRef]
4. Fletcher, C.D.M. The Evolving Classification of Soft Tissue Tumours—An Update Based on the New 2013 WHO Classification. *Histopathology* **2014**, *64*, 2–11. [CrossRef] [PubMed]
5. Frith, A.E.; Hirbe, A.C.; Van Tine, B.A. Novel Pathways and Molecular Targets for the Treatment of Sarcoma. *Curr. Oncol. Rep.* **2013**, *15*, 378–385. [CrossRef] [PubMed]
6. Demicco, E.G.; Park, M.S.; Araujo, D.M.; Fox, P.S.; Bassett, R.L.; Pollock, R.E.; Lazar, A.J.; Wang, W.L. Solitary Fibrous Tumor: A Clinicopathological Study of 110 Cases and Proposed Risk Assessment Model. *Mod. Pathol.* **2012**, *25*, 1298–1306. [CrossRef]
7. England, D.M.; Hochholzer, L.; McCarthy, M.J. Localized Benign and Malignant Fibrous Tumors of the Pleura. A Clinicopathologic Review of 223 Cases. *Am. J. Surg. Pathol.* **1989**, *13*, 640–658. [CrossRef]
8. Hassani, M.; Jung, S.; Ghodsi, E.; Seddigh, L.; Kooner, P.; Aoude, A.; Turcotte, R. Value of Cellular Components and Focal Dedifferentiation to Predict the Risk of Metastasis in a Benign-Appearing Extra-Meningeal Solitary Fibrous Tumor: An Original Series from a Tertiary Sarcoma Center. *Cancers* **2023**, *15*, 1441. [CrossRef]
9. Tolstrup, J.; Loya, A.; Aggerholm-Pedersen, N.; Preisler, L.; Penninga, L. Risk Factors for Recurrent Disease after Resection of Solitary Fibrous Tumor: A Systematic Review. *Front. Surg.* **2024**, *11*, 1–11. [CrossRef]
10. Medina-Ceballos, E.; Machado, I.; Giner, F.; Bujeda, Á.B.; Navarro, S.; Ferrandez, A.; Lavernia, J.; Ruíz-Sauri, A.; Llombart-Bosch, A. Solitary Fibrous Tumor: Can the New Huang Risk Stratification System for Orbital Tumors Improve Prognostic Accuracy in Other Tumor Locations? *Pathol. Res. Pract.* **2024**, *254*, 6–11. [CrossRef]
11. Salas, S.; Resseguier, N.; Blay, J.Y.; Le Cesne, A.; Italiano, A.; Chevreau, C.; Rosset, P.; Isambert, N.; Soulie, P.; Cupissol, D.; et al. Prediction of Local and Metastatic Recurrence in Solitary Fibrous Tumor: Construction of a Risk Calculator in a Multicenter Cohort from the French Sarcoma Group (FSG) Database. *Ann. Oncol.* **2017**, *28*, 1979–1987. [CrossRef] [PubMed]
12. Gholami, S.; Cassidy, M.R.; Kirane, A.; Kuk, D.; Zanchelli, B.; Antonescu, C.R.; Singer, S.; Brennan, M. Size and Location Are the Most Important Risk Factors for Malignant Behavior in Resected Solitary Fibrous Tumors. *Ann. Surg. Oncol.* **2017**, *24*, 3865–3871. [CrossRef] [PubMed]
13. Pasquali, S.; Gronchi, A.; Strauss, D.; Bonvalot, S.; Jeys, L.; Stacchiotti, S.; Hayes, A.; Honore, C.; Collini, P.; Renne, S.L.; et al. Resectable Extra-Pleural and Extra-Meningeal Solitary Fibrous Tumours: A Multi-Centre Prognostic Study. *Eur. J. Surg. Oncol.* **2016**, *42*, 1064–1070. [CrossRef] [PubMed]
14. Wushou, A.; Jiang, Y.Z.; Liu, Y.R.; Shao, Z.M. The Demographic Features, Clinicopathologic Characteristics, Treatment Outcome and Disease-Specific Prognostic Factors of Solitary Fibrous Tumor: A Population-Based Analysis. *Oncotarget* **2015**, *6*, 41875–41883. [CrossRef] [PubMed]
15. Wang, K.; Mei, F.; Wu, S.; Tan, Z. Hemangiopericytoma: Incidence, Treatment, and Prognosis Analysis Based on SEER Database. *Biomed. Res. Int.* **2020**, *2020*, 2468320. [CrossRef]
16. Yamada, Y.; Kohashi, K.; Kinoshita, I.; Yamamoto, H.; Iwasaki, T.; Yoshimoto, M.; Ishihara, S.; Toda, Y.; Itou, Y.; Koga, Y.; et al. Clinicopathological Review of Solitary Fibrous Tumors: Dedifferentiation Is a Major Cause of Patient Death. *Virchows Arch.* **2019**, *475*, 467–477. [CrossRef]
17. Hall, W.A.; Ali, A.N.; Gullett, N.; Crocker, I.; Landry, J.C.; Shu, H.K.; Prabhu, R.; Curran, W. Comparing Central Nervous System (CNS) and Extra-CNS Hemangiopericytomas in the Surveillance, Epidemiology, and End Results Program: Analysis of 655 Patients and Review of Current Literature. *Cancer* **2012**, *118*, 5331–5338. [CrossRef]
18. Howlader, N.; Noone, A.; Krapcho, M.; Miller, D.; Bishop, K.; Altekruse, S.; Kosary, C.; Yu, M.; Ruhl, J.; Tatalovich, Z.; et al. SEER Cancer Statistics Review. Available online: https://seer.cancer.gov/csr/1975_2017/ (accessed on 1 September 2024).
19. Haas, R.L.; Walraven, I.; Lecointe-Artzner, E.; van Houdt, W.J.; Strauss, D.; Schrage, Y.; Hayes, A.J.; Raut, C.P.; Fairweather, M.; Baldini, E.H.; et al. Extrameningeal Solitary Fibrous Tumors—Surgery Alone or Surgery plus Perioperative Radiotherapy: A Retrospective Study from the Global Solitary Fibrous Tumor Initiative in Collaboration with the Sarcoma Patients EuroNet. *Cancer* **2020**, *126*, 3002–3012. [CrossRef]
20. Localized/Regional/Distant Stage Adjustments. Available online: https://seer.cancer.gov/seerstat/variables/seer/yr1975_2017/lrd_stage/index.html (accessed on 15 January 2023).
21. Wilson, A.; Norden, N. The R Project for Statistical Computing the R Project for Statistical Computing. Available online: https://www.r-project.org/ (accessed on 1 September 2024).
22. Georgiesh, T.; Boye, K.; Bjerkehagen, B. A Novel Risk Score to Predict Early and Late Recurrence in Solitary Fibrous Tumour. *Histopathology* **2020**, *77*, 123–132. [CrossRef]
23. Demicco, E.G.; Griffin, A.M.; Gladdy, R.A.; Dickson, B.C.; Ferguson, P.C.; Swallow, C.J.; Wunder, J.S.; Wang, W.L. Comparison of Published Risk Models for Prediction of Outcome in Patients with Extrameningeal Solitary Fibrous Tumour. *Histopathology* **2019**, *75*, 723–737. [CrossRef]
24. O'Neill, A.C.; Tirumani, S.H.; Do, W.S.; Keraliya, A.R.; Hornick, J.L.; Shinagare, A.B.; Ramaiya, N.H. Metastatic Patterns of Solitary Fibrous Tumors: A Single-Institution Experience. *Am. J. Roentgenol.* **2017**, *208*, 2–9. [CrossRef] [PubMed]

25. Luzzago, S.; Palumbo, C.; Rosiello, G.; Knipper, S.; Pecoraro, A.; Mistretta, F.A.; Tian, Z.; Musi, G.; Montanari, E.; Soulières, D.; et al. Association Between Systemic Therapy and/or Cytoreductive Nephrectomy and Survival in Contemporary Metastatic Non–Clear Cell Renal Cell Carcinoma Patients. *Eur. Urol. Focus.* **2021**, *7*, 598–607. [CrossRef] [PubMed]
26. Würnschimmel, C.; Wenzel, M.; Collà Ruvolo, C.; Nocera, L.; Tian, Z.; Saad, F.; Briganti, A.; Shariat, S.F.; Mirone, V.; Chun, F.K.H.; et al. Life Expectancy in Metastatic Prostate Cancer Patients According to Racial/Ethnic Groups. *Int. J. Urol.* **2021**, *28*, 862–869. [CrossRef] [PubMed]
27. Krengli, M.; Cena, T.; Zilli, T.; Jereczek-Fossa, B.A.; De Bari, B.; Villa Freixa, S.; Kaanders, J.H.A.M.; Torrente, S.; Pasquier, D.; Sole, C.V.; et al. Radiotherapy in the Treatment of Extracranial Hemangiopericytoma/Solitary Fibrous Tumor: Study from the Rare Cancer Network. *Radiother. Oncol.* **2020**, *144*, 114–120. [CrossRef]
28. Bishop, A.J.; Zagars, G.K.; Demicco, E.G.; Wang, W.L.; Feig, B.W.; Guadagnolo, B.A. Soft Tissue Solitary Fibrous Tumor Combined Surgery and Radiation Therapy Results in Excellent Local Control. *Am. J. Clin. Oncol.* **2018**, *41*, 81–85. [CrossRef]
29. Vaz Salgado, M.A.; Soto, M.; Reguero, M.E.; Muñoz, G.; Cabañero, A.; Gallego, I.; Resano, S.; Longo, F.; Madariaga, A.; Gomez, A.; et al. Clinical Behavior of Solitary Fibrous Tumor: A Retrospective Review of 30 Patients. *Clin. Transl. Oncol.* **2017**, *19*, 357–363. [CrossRef]
30. Demicco, E.G.; Wagner, M.J.; Maki, R.G.; Gupta, V.; Iofin, I.; Lazar, A.J.; Wang, W.L. Risk Assessment in Solitary Fibrous Tumors: Validation and Refinement of a Risk Stratification Model. *Mod. Pathol.* **2017**, *30*, 1433–1442. [CrossRef]
31. Colditz, G.A. American Joint Committee on Cancer. In *The SAGE Encyclopedia of Cancer and Society*; Springer: New York, NY, USA, 2015; pp. 1–2.

Review

Surgery for Solitary Fibrous Tumors of the Pleura: A Review of the Available Evidence

Pietro Bertoglio [1,2,*], Giulia Querzoli [3], Peter Kestenholz [4], Marco Scarci [5], Marilina La Porta [1], Piergiorgio Solli [1] and Fabrizio Minervini [4]

1   Division of Thoracic Surgery, IRCCS Azienda Ospedaliero Universitaria di Bologna, 40100 Bologna, Italy; marilina.laporta@studio.unibo.it (M.L.P.); piergiorgio.solli@gmail.com (P.S.)
2   Alma Mater Studiorum, University of Bologna, 40064 Bologna, Italy
3   Pathology Unit, IRCCS Azienda Ospedaliero Universitaria di Bologna, 40100 Bologna, Italy; querzoligiulia@gmail.com
4   Division of Thoracic Surgery, Cantonal Hospital of Lucerne, 6000 Lucerne, Switzerland; peter.kestenholz@luks.ch (P.K.); fabriziominervini@hotmail.com (F.M.)
5   Department of Thoracic Surgery, Imperial College Healthcare NHS Trust, London W2 1NY, UK; marco.scarci@mac.com
*   Correspondence: pieberto@hotmail.com; Tel.: +39-0516478111

Simple Summary: This is a review of the evidence on the use of surgery for solitary fibrous tumors of the pleura. Solitary fibrous tumors of the pleura are rare tumors that can arise in the chest from both visceral and parietal pleura. Surgery is the cornerstone of their treatment; as minimally invasive techniques, both thoracoscopy or robotics can be used according to the dimension, position, and infiltration of neighboring organs. A radical resection with free margins is the main target of surgery. Even if the long-term results are generally good, the risk of local or distant recurrence is always possible, in particular in cases of more aggressive histological types. In cases of local recurrence, surgery can be proposed if feasible.

Abstract: Solitary fibrous tumors of the pleura (pSFT) are a relatively rare neoplasms that can arise from either visceral or parietal pleura and may have different aggressive biological behaviors. Surgery is well known to be the cornerstone of the treatment for pSFT. We reviewed the existing literature, focusing on the role of surgery in the management and treatment of pSFT. All English-written literature has been reviewed, focusing on those reporting on the perioperative management and postoperative outcomes. Surgery for pSFT is feasible and safe in all experiences reported in the literature, but surgical approaches and techniques may vary according to the tumor dimensions, localization, and surgeons' skills. Long-term outcomes are good, with a 10-year overall survival rate of more than 70% in most of the reported experiences; on the other hand, recurrence may happen in up to 17% of cases, which occurs mainly in the first two years after surgery, but case reports suggest the need for a longer follow-up to assess the risk of late recurrence. Malignant histology and dimensions are the most recognized risk factors for recurrence. Recurrence might be operated on in select patients. Surgery is the treatment of choice in pSFT, but a radical resection and a careful postoperative follow-up should be carried out.

Keywords: pleural solitary fibrous tumors; surgery; minimally invasive surgery; overall survival

Citation: Bertoglio, P.; Querzoli, G.; Kestenholz, P.; Scarci, M.; La Porta, M.; Solli, P.; Minervini, F. Surgery for Solitary Fibrous Tumors of the Pleura: A Review of the Available Evidence. *Cancers* **2023**, *15*, 4166. https://doi.org/10.3390/cancers15164166

Academic Editor: Bahil Ghanim

Received: 31 July 2023
Revised: 12 August 2023
Accepted: 13 August 2023
Published: 18 August 2023

1. Introduction

Solitary fibrous tumors of the pleura (pSFT) are rare neoplasms arising from either visceral or parietal pleura, accounting for 4% of chest tumors with an incidence of 2.8 per 100,000 [1]. In the past decades, it was commonly known as "benign pleural mesothelioma" [2]. pSFT can occur at any age but is more frequently observed in middle-aged patients with a peak of incidence in the fourth and sixth decade (median in the fifth decade), with no gender predilection; moreover, some isolated cases in children and adolescents have been occasionally reported [3,4]. From a pathological point of view, according to the 2020 World Health Organization Classification, SFTs are classified as mesenchymal tumors of fibroblastic or myofibroblastic origin with intermediate biological behavior [2].

Although the most common site of onset is the pleura, similar mesenchymal tumors can arise anywhere in the body, such as the head and neck, thoracic wall, mediastinum, pericardium, retroperitoneum, and abdominal cavity; another described location includes the central nervous system, such as the meninges and spinal cord [5,6].

Biological behavior might vary from a benign tumor with an indolent course and a slow growth rate to a more aggressive tumor with the development of distant metastases [4]. Possible clinical manifestations are generally related to large masses and the most common symptoms are dyspnea, cough, pleuritic chest pain, shortness of breath, fever, and weight loss [3].

To date, the gold standard in the treatment of pSFT remains surgery, with the aim of a radical (R0) resection.

This review aims to analyze the role and the outcomes of surgery in the treatment of pSFT.

2. Methods

Given the rarity of pSFT, no prospective or randomized studies are available in the literature. For this reason, selected sources included case reports and observational and experimental studies of patients undergoing surgery for the diagnosis of pSFT.

Search Strategy

We searched PubMed databases for RCTs, observational studies (prospective or retrospective), and case reports through to May 2023. PubMed was searched with the following keywords: "solitary fibrous tumor pleural" and "classification", or "treatment", or "surgery", or "minimally invasive surgery", or "outcome", or "recurrence", or "survival", or "prognostic factors". Only manuscripts with the full text in English were considered.

3. Histology and Molecular Features

From a histological point of view, SFT can be divided in two main phenotypes: hypercellular and hypocellular. pSFT are represented by a hypocellular phenotype, that is characterized by a thick collagenous background, often associated to hyalinized or collagen bands. Within the stroma, cells have a spindle shape, may be atypical, and they are arrayed haphazardly in a storiform configuration or in randomly oriented fascicle ("patternless pattern") [2]. Areas with higher and lower cellularity usually alternate. On the other hand, the hypercellular type of SFTs are poor in collagen fibers and several solid nests are separated by capillaries. In this type of SFT, the cell shape is usually ovoid. A hemangiopericytic and perivascular pattern are the most characteristic histological patterns for SFT. Hemorrhage is common in cellular tumors, and necrosis may be present, particularly in malignant histotypes.

STAT6 is the most useful diagnostic marker, and it is expressed in 95% of cases. Other immunohistochemistry markers of SFT include positivity for CD34, Bcl2, CD99, and vimentin, while actin, desmin, S100 protein, or epithelial markers are negative. Nevertheless, there is a variability in the expression of the aforementioned markers.

From a molecular point of view, SFT typically show inversion of the long arm of chromosome 12 that brings to the fusion of the genes NAB2 and STAT6, which is typical of the macrofamily of SFT [7–9]. The NAB2 gene function is eventually involved in cellular differentiation and proliferation, while STAT6 acts as a transcriptional transactivator. The fusion of the two aforementioned genes is the driver of the tumorigenesis [8]. Interestingly, insulin-like growth factor-2 (IGF-2), which is implicated in the pathogenesis of Doege–Potter syndrome, one of the paraneoplastic syndromes that may be associated with pSFT, is dysregulated by the NAB2/STAT6 mutation [8].

Nonetheless, no molecular characteristics that are able to discriminate and stratify malignant and benign pSFTs are available so far. As a matter of fact, it has been suggested that some specific NAB2–STAT6 fusion could be related to a more aggressive biological behavior [10], but this finding has been questioned by other studies [11,12]

On the other hand, in up to 30% of all SFTs, a mutation in the promoter of the telomerase reverse transcriptase has been associated with worse prognostic outcomes and a more aggressive behavior [13–15], although the results of other studies are not consistent [16].

4. Classification

In the latest WHO classification (5th edition), the pSFT description encompasses various ICD-O codes (8815/0 solitary fibrous tumor, benign, 8815/1 solitary fibrous tumor NOS, 8815/3 solitary fibrous tumor, malignant). The effective prognostic role of conventional staging systems, such as the tumor, node, metastasis (TNM) staging system, is still not clear. The American Joint Committee on Cancer (AJCC)/Union for International Cancer Control (UICC) suggest to consider malignant pSFTs as soft tissue sarcomas, even though a clear and standardized definition of malignancies has not been released by these authorities [2].

For these reasons, many authors preferred to refer to a risk stratification model rather than anatomical staging. To better stratify clinical behavior and the subsequent risk of recurrence, different classification systems of pSFT have been proposed. All the classification systems are based on pathological or macroscopic features of the pSFT, and they aim to predict prognosis and the risk of recurrence. Due to the rarity of the tumors, the main bias of all the classification systems is to be based on the relatively small retrospective and often monocentric series.

Table 1 reports the characteristics off all of the proposed classification systems. All of them are based on histopathological characteristics: high cellularity with crowding and overlapping of nuclei, cellular pleomorphism, high mitotic index (with a general cut-off at four per high-power fields), necrosis or stromal/vascular invasion.

The first classification system was proposed by England [4] in 1989 and it was mainly based on histological features taking into account cellularity, mitosis, pleomorphism, necrosis, and presence of hemorrhage; dimension and other macroscopic feature were taken into account, but only the presence of one or more microscopic characteristics were related to the diagnosis of high risk ("malignant") pSFT. In 2002, De Perrot and his coworkers [17] proposed to divide pSFT in four stages with a progressive correlation with worse outcomes according to macroscopic features and pathological characteristics. Together with the histologic features reported by England, the authors associate the presence of a pedunculated (lower risk) or sessile (higher risk) lesions.

The real prognostic impact of the latter classification has been discussed. Two retrospective series analyzed the impact of the De Perrot classification in their own institutional dataset of malignant pSFT. In both manuscripts, the authors failed to verify the prognostic relevance of this classification for overall survival [18,19]; on the other hand, Schirosi and colleagues [20] confirm the prognostic value of this classification.

More recently, in 2013, Tapias et al. [21] tried to standardize a classification and a score system to identify a reliable prediction of relapse that could help to address a long and appropriate follow-up only for patients with a high risk of recurrence. The scoring system was established, assigning one point to each of six tumor features. The scoring system was found to be predictive for tumor relapse when a score of ≥ 3 points was used as the cut-off, distinguishing low-risk patients from high-risk ones (the number of the patients for the validation of the score was too low to define an intermediate risk cohort). The same authors performed a follow-up study that, on one hand, showed a lower prognostic impact compared to the first study, but, on the other hand, it still had higher prognostic potential compared to [22].

In addition, Diebold et al. [23] conducted a retrospective analysis on 78 patients and reported a mean overall survival after surgery of 172 ± 13 months and a mean event-free survival of 165 ($\pm$15) months, with a median follow-up time of 36 months (range 1–216). Six patients (7.7%) had an adverse outcome, including pSFT relapse. Relying on these data, they proposed a scoring system expanding those from De Perrot [17] and Tapias [21] and introducing a new variable, namely the proliferation marker MIB-1.

Table 1. Table 1 reports all the elements included in different classification systems.

	England [4]	De Perrot [17]	Tapias [21]	Diebold [23]	Demicco [6]
Histologic features for malignancy					
Mitotic figures per 10 HPF	>4	>4	$\geq$4	$\geq$4	0 1–3 >4
Hypercellularity	Yes	Yes	Yes		
Pleomorphism	Yes	Yes			
Necrosis	Yes	Yes	Yes	Yes	Yes
Hemorrhage	Yes		Yes		
Stromal/vascular invasion		Yes			
Anatomic feature for malignancy					
Size	>10 cm		$\geq$10 cm	$\geq$10 cm	<5 5–10 10–15 $\geq$15
Peduncolated/sessile	Sessile	Sessile	Sessile		
Site	Parietal pleura		Parietal pleura		
Biomarker risk factors					
MIB-1 proliferation index				$\geq$10%	
Patients related risk factors					
Age					$\geq$55
Criteria for malignancy					
	One or more histologic criteria	One or more histologic criteria	Any three histologic and anatomic criteria	Two or more histologic, anatomic, or biomarker criteria	Low, moderate, or high risk according to the final score

Lastly, in 2017, Demicco [24] proposed another risk prediction model including patient age, tumor size, and mitotic activity; they then also introduced necrosis. This risk prediction model was not intended only for pSFT of the pleura and could be used also for extra-thoracic pSFT.

Several studies compared the different classification models; in their follow up study, as already mentioned, Tapias [22] showed a significantly better recurrence prediction of their model compared to that of England [4] and De Perrot [17]. Similarly, Ricciardi [18] found a significantly better prognostic role for the Tapias score compared to the De Perrot and Demicco, while Bellini [25] confirmed the prognostic impact of the England, Tapias, and Diebold scores, but not that by De Perrot.

5. Clinical Presentation

The vast majority of patients have no clear or specific symptoms at diagnosis; in some cases, intrathoracic masses, even of large dimensions, can be found incidentally at the time of the imaging test, which are carried out for other reasons with no symptoms reported. Regarding symptoms of dyspnea, cough or chest pain are the most typically reported [3,22,26,27]. Hemoptysis or pneumonitis has been described in case of large masses that cause atelectasis of the lung parenchyma [17,22].

Given the slow growth, pSFTs can reach impressive dimensions with poor or even no symptoms. The compression of large vessels or of the heart are the ultimate cause of signs that require further investigation and consequent diagnosis.

In rare cases, pSFT has been reported to be associated with paraneoplastic syndromes such as digital clubbing and hypertrophic pulmonary osteoarthropathy (HPO, Pierre–Marie–

Bamberger syndrome); no clear mechanism bringing to the development of this syndrome has been found, which could be related to the secretion of cytokines and hyaluronic acid by the tumor and chronic hypoxia [28–30].

Furthermore, Doege–Potter syndrome has also been related to pSFT; this syndrome is characterized by a refractory hypoglycemia and involves less than 5% of all SFTs, even if it seems to be more common in large pleural or peritoneal lesions [31–33]. This syndrome is related to the secretion of insulin-like growth factor 2 (IGF-2) [34,35].

6. Surgical Management in the Treatment of pSFT

6.1. Preoperative Assessment

As discussed in the previous paragraphs, a diagnosis of pSFT is usually due to unexpected findings in chest radiological images or due to symptoms related to the dimension of the tumor.

Preoperative evaluation usually requires only a CT scan or, according to the surgeon's preference, MRI. The pSFT does not have pathognomonic features and the characteristics are similar to those of other types of soft-tissue neoplasm. In more detail, pSFTs usually appear as homogeneous masses and are well defined; larger tumors may contain calcifications or cystic areas [36,37]. The neighboring organs are more frequently displaced rather than infiltrated [38,39]. Using MRI, pSFTs show low intensity T1 signaling and variable T2 signaling. The presence of a pedicle (which is reported in roughly 40% of cases) might result in a change in the location and shape of the pSFT [40,41]. In some cases, pleural effusion might be present [4,42].

In the case of small or well delineated lesions, MRI does not add further information compared to CT scanning, but it might be useful in cases of suspicious chest wall, vertebral, or diaphragmatic infiltration [27,38].

Beyond radiological tests, further exams, such as bronchoscopy, might be necessary in selected cases [30]. According to the extension of resection needed, pulmonary function tests might be necessary. Preoperative biopsy is not mandatory as the sensitivity has been reported to be low [43]; it might be performed in selected cases of unresectable advanced diseases or if a diagnostic doubt is present. Moreover, fine or core needle biopsies have a low sensitivity and malignant features are hard to identify with a small specimen [44].

Concurrently, [18]FDG-PET scanning is not routinely required in the preoperative assessment of pSFTs [45]. Recently, a single-institutional study suggested [46] that [18]FDG-PET scanning might have a role in stratifying the clinical behavior of pSFTs, identifying those with a more aggressive or malignant component. Yeom and colleagues [47] retrospectively reviewed preoperative CT and [18]FDG-PET images of pathologically proven pSFTs and reported that morphological and metabolic features might help clinicians in the differential diagnosis; conversely, they could not predict the biological behavior of these neoplasms, but found a significantly higher standard uptake value (SUV) in patients with a malignant pSFT.

Lococo [48] reported the experimental use of ^{68}Ga-Dotatoc in the preoperative workup of pSFTs, but only anecdotical further experiences using this technique has been reported so far [45], and, although results seem promising, no clear indications regarding its predictive and prognostic role are available to date.

In case of large masses, some authors suggest preoperative embolization of arterial vessels in order to reduce the risk of hemorrhage and minimize intraoperative blood loss [49,50]. It has been reported that up to 50% of pSFTs have a vascular pedicle that might arise form intercostal, internal mammary, or bronchial arteries [51,52]. In addition, abdominal feeding vessels from the abdominal aorta or celiac tripod have been reported [53–55]. In a case report, Song [56] reported the surgical ligation of the feeding vessel via an anterior thoracotomy followed by pSFT resection through a lateral thoracotomy. Nevertheless, no standardized indications for preoperative embolization are available and the decision is based on the multidisciplinary evaluation of each case.

6.2. Operative Approach

As in most oncological surgeries, the principles of pSFT surgery require a microscopically radical (R0) en bloc resection with a free safe margin [4,6,57]. Additional hilar or mediastinal lymphadenectomy is not recommended unless a clinical suspect is present [27,28,58]. As already mentioned, to date, the surgical approach represents the cornerstone of pSFT treatment and should be considered whenever possible. Yao and colleagues [59] analyzed data from the National Cancer DataBase (NCDB) of patients treated (either surgery or with other treatments) between 2004 and 2014 for pSFT; the authors found that the outcomes of those who received surgery were significantly better in terms of overall survival (64% versus 22% at 5 years).

The extent of surgical resection should vary according to the characteristics and features of pSFTs. When the tumor is pedunculated and it arises from the visceral pleura, a wedge resection often guarantees a radical resection with an adequate free margin; on the other hand, sessile tumors might require a wider parenchymal resection, which might range from a wedge resection to even a pneumonectomy in cases of larger or giant tumors [60]. Concurrently, in cases of pSFT arising from the parietal pleura, an extra-pleural preparation and resection is the gold standard with the addition of chest wall or spine resection if the tumor is invading chest wall structures. In case of doubt of radicality, a frozen section can be required during the surgical procedure [61]. In a relatively large series from a single center, Lahon and colleagues [27] reported 65% wedge resections, 10% lobectomies, and 3% pneumonectomy; interestingly, roughly 8% of surgeries required extended resections (chest wall, diaphragm, or vertebrae). On the other hand, more recently and based on the data of NCDB, Yao [59] found an almost equal rate of wedge resections and lobectomies (42% versus 45% respectively), but a higher rate of pneumonectomies (13%); the authors did not find a difference in OS when they compared lobectomies and wedge resections.

Different surgical approaches have been described, according not only to the period and the availability of technologies, but also, especially in recent times, to the dimension and the organ infiltration. Open surgery has been considered the gold standard approach; thoracotomy, sternotomy, or even hemi or complete clamshell, have been described in the literature [44,62,63]. Posterolateral was reported as the preferred access in different series [29,30,64] in cases of larger tumors.

In minimally invasive surgery, both VATS and RATS have been gradually introduced, given the known advantages in reducing postoperative pain and in-hospital stay. Minimally invasive surgery, even using a single port, can be offered in case of tumors of limited dimension, which allow safe and proper intraoperative manipulation, but the indication for the use of a minimally invasive technique is strictly related to the skills and knowledge of surgeons [44,65–69]. In a recent Chinese report [70], the authors found that patients treated with VATS had a significantly smaller diameter and a faster postoperative course compared to those treated with thoracotomy. Cardillo reported a conversion rate of 14.5% from VATS to open [71]; it must be noted that over time, the use of minimally invasive surgery has largely developed, and the conversion rate might have dramatically reduced. Nonetheless, patient safety and radical resection should be the priority for possible conversion from minimally invasive to open surgery.

Removal of the specimen in VATS and RATS must be performed with attention and by using a bag or by covering the utility incision with a soft tissue retractor in order to avoid possible contact metastases, which has been described by some authors [58]. In cases of large pSFTs, in order to avoid an enlargement of the minimally invasive surgical access and consequent injuries to the intercostal nerves, Hatooka [72] suggests use of a subxiphoid access.

6.3. Operative Management

Preoperative careful evaluation of imaging is mandatory for a safe surgical resection. As a matter of fact, intraoperative management strictly relies on the tumor dimensions and possible infiltration of the surrounding organs.

According to the tumor features, a 45° lateral decubitus position can be of extreme help in avoiding hemodynamic variations associated with decompressing the heart or great vessels [44]. A standard minimally invasive or open approach can be used in case of small or easily reachable tumors, while chest wall, diaphragm, or mediastinal structure invasion might require a dedicated surgical approach.

Intraoperative management should be customized in order to have the best exposure and to reach a R0 resection [66].

Intraoperative mortality in pSFT resection has been reported to range between 1.5% and 12% [44]; as already discussed, in case of large masses, the first step should be recognition and ligation of the feeding vessels in order to reduce bleeding and reduce intraoperative hemorrhage.

In cases of giant tumors (>20 cm) in the literature, an open approach is usually the treatment of choice. Association of different approaches, such as multi-level thoracotomies or thoracotomy and subcostal access, might be necessary to radically and safely resect giant pSFTs [56,73–75].

6.4. Postoperative Outcomes

The postoperative outcomes and the complication rate after surgery for pSFTs strictly depend on the extent of surgical resection and, quite obviously, on the dimension of the tumor. Bleeding and respiratory distress are the most reported complications. As reported before, particularly in cases of large tumors, preoperative embolization might help to prevent or reduce intraoperative bleeding [49–52].

In more detail, Lahon and coworkers [27] reported a single case of 30-day mortality and 5.7% of morbidity among 157 patients. Interestingly, almost 80% of patients that developed a postoperative complication had a malignant histotype; acute respiratory distress syndrome was the most common complication, while hemothorax was found in a single patient. The mean reported length of hospital stay was 11.5 ± 4 days. On the other hand, Tapias [21] reported a shorter length of stay (median 3 days, range 1–15), but they had a higher rate of minimally invasive surgery. Concurrently, Bellini [25] found a significant correlation between the presence of a giant tumor and intraoperative bleeding and a trend towards a longer in-hospital stay. In addition to this, in their manuscript focused on malignant pSFTs, Lococo and colleagues [26] highlight a 26% rate of postoperative complications (the most common were atrial fibrillation, bleeding, and pneumonia) and two cases of postoperative mortality; the first patient received an intrapericardial left pneumonectomy and the second one a lobectomy associated to a chest wall resection. Similar complications were also reported by a report from the Mayo Clinic, which additionally highlighted a 3.6% 30-day postoperative mortality rate [43].

6.5. Neoadjuvant and Adjuvant Treatments

Preoperative therapies have been seldom reported with no clear indications and, to date, are not recommended [17,76]. Out of 110 patients with SFT from several sites, Demicco [24] reported 9 cases who underwent preoperative chemo and/or radiotherapy. The possible influence of preoperative treatment on the long-term outcomes was not reported.

Similarly, the role of adjuvant therapies is still under debate as their real benefit has not been studied in large prospective studies, and results are often inconsistent [77–80].

Adjuvant radiotherapy might be suggested in patients with histological risk factors (e.g., higher mitotic count) or R1/R2 resections [24]. Suter and colleagues [81], in their own institutional series, reported a single case of adjuvant radiotherapy with excellent long-term outcomes. On the other hand, Cardillo reported a local recurrence [28] after chest wall resection and adjuvant radiotherapy (30 Gy) for a malignant pSFT. Adjuvant radiotherapy was also offered to 18 patients (20.9%) in a large retrospective study [79], which also involved extra-thoracic SFTs. The role of radiotherapy in the treatment of pSFTs has been also evaluated in a retrospective study based on 40 patients [82], that reported good long-term outcomes after definitive radiotherapy both in a curative (60 Gy) and in a palliative (39 Gy) setting (5 year OS 87.5% and 54.2%, respectively). It must be underlined that among the whole cohort, only nine patients had a pleural SFT (one in the definitive treatment setting and eight in the palliative setting). Radiotherapy has been anecdotally used for recurrence with promising results [80,83].

Similar to radiotherapy, the role of adjuvant chemotherapy has not been fully cleared and it is not standardized [17]. Chemotherapy has been described mainly for recurrence and the most common drugs that were reported are doxorubicin-based and gemcitabine-based [84]. More recently, research focused on the possible use of angiogenetic drugs, such as sunitinib, pazopanib, temozolomide, and bevacizumab [85–88]. In an in vitro study, Ghanim [77] reported the potential use of Trabectidin in association with Ponatinib. Despite the promising results, the data is currently not sufficient to establish new standards of care.

7. Long Term Results

7.1. Overall Survival

Overall survival after resection of pSFT is generally good. Table 2 reports main results of the largest and most recent study. Lahon and colleagues [27] reported a median OS of 14 years, a 5 year OS rate of 86%, and a 10 years OS of 77%; survival was significantly impaired in patients with a malignant pSFTs compared to those with a benign disease (5 year OS, 68% and 96%, respectively). Similar conclusions were shared by several other authors; Lococo and colleagues [26] reported an OS rate of 81% and 67% for benign and malignant pSFTs, respectively. Concurrently, in a more recent report with a median follow up of 91 months, Zhou and coworkers [63] did not report any death in the benign pSFT group and they found a 94.5 year OS, with a significant difference between benign and malignant disease. Similarly, Cardillo and his group [71] analyzed the outcomes of 110 patients who underwent surgery for pSFT and found a 10 year OS of 97.5. Lastly, in a small observational cohort study, Franzen and colleagues [89] found a pSFT-related mortality of 7.1%.

According to the majority of reported series, radicality of surgical excision (R0 resection) along with tumor size (larger or smaller than 10 cm), mitotic rate, presence of necrosis, and/or hemorrhage are well-recognized prognostic factors for pSFT [23].

7.2. Disease-Free Survival, Recurrence, and Recurrence Treatment

Table 3 reports features of recurrence. The recurrence rate for pSFT has been reported to be between 5% and 17%, but it grows up to an interval between 14 and 54% in case of malignant pSFTs [3,26,30,43,71]. Time to recurrence might be very long and it has been reported in case reports to be up to 17 years after surgery [90], but the majority occur in the first 2 years [76]. For this reason, both recurrence treatment and the correct follow-up period are not clear based on the current guidelines. The NCCN guidelines [91] include pSFT in the group of soft tissue sarcoma and recommend a very strict follow-up according to surgical results. Given the higher recurrence rate, more careful postoperative surveillance is required for malignant pSFTs with more strict controls and for a longer time compared to benign pSFTs [63]. The majority of the authors suggest a CT scan every 6 months for the first 2 years following surgical resection, then a yearly CT scan afterwards; PET scans may have a role in the follow-up if the first pSFT was PET positive [17,44].

A recent meta-analysis published in 202 based on 23 retrospective studies [92] found a 9% recurrence rate after surgery (even in case of R0 resections); the recurrence rate was significantly different according to the pathological features of the pSFT (benign and malignant: 3% and 22% recurrence rate, respectively); nevertheless, no significant difference was seen according to the site of origin of the tumor (visceral or parietal pleura). The impact of malignant histology on DFS has been confirmed by several authors [25,27,63,93,94]. On the other hand, Bellini and her colleagues [25] found a significantly worse DFS in patients who had a pSFT arising from parietal pleura. Moreover, in a multi-centric international study, Ghanim focused on circulating biomarkers as prognostic factors; the authors found that higher levels of fibrinogen were related with a significantly lower DFS, while the neutrophil–lymphocyte ratio (NLR) had no influence on DFS. Franzen [89] reported that the tumor diameter, the number of mitotic cells, and Ki67 expression as prognostic factors significantly influencing the DFS. Relapse might be systemic or, more frequently, local, close to resection margin. In case of local relapse, surgery can still be offered to patients if technically feasible and is usually considered the treatment of choice [17]. In their manuscript, Lahon and colleagues [27] reported re-surgery in 9 of the 10 cases with local recurrence; similarly, Lococo and his coworkers [26] reported an addition surgical resection in 4 patients out of 6 with local recurrence.

Table 2. Main results of recent studies reporting the outcomes of pSFTs treated with surgery. OS: Overall Survival; DFS: Disease-Free Survival; VATS: Video Assisted Thoracic Surgery; RATS: Robot Assisted Thoracic Surgery.

Authors and Year	Monocentric/ Multi-centric	Number of Patients	Benign/Malignant	Type of Surgery	Open/Minimally Invasive	5 Year OS	10 Year OS	5 Year DFS	10 Year DFS	Follow-Up
Magdeilenat et al. 2002 [30]	Multicentric	60	38 benign 22 malignant	42 wedge resection 7 parietal pleura excision 11 extended resection	54 open (48 posterolateral, 5 anterolateral, 1 median sternotomy) 6 VATS	94%	94%	NA	NA	88 months (mean)
Harrison-Phipps et al. 2009 [43]	Monocentric	84	73 benign 11 malignant	62 wedge resection 2 segmentectomy 4 lobectomy 7 parietal pleura excision 9 extended resection	72 open (68 posterolateral thoracotomy, 3 median sternotomy, 1 transabdominal approach) 12 VATS	83%	NA	95%	NA	12 years (median)
Cardillo et al. 2009 [71]	Monocentric	110	95 benign 15 malignant	88 wedge resection 6 lobectomy 1 bilobectomy 2 pneumonectomy 12 parietal pleura excision 2 extended resection	41 open (40 thoracotomy, 1 median sternotomy) 69 VATS (16 conversion to open	NA	97.5%	NA	90.8%	12–222 months (range)
Lahon et al. 2012 [27]	Monocentric	157	90 benign 67 malignant	122 wedge resection 1 segmentectomy 15 lobectomy 3 bilobectomy 4 pneumonectomy 12 extended resection	142 open (139 posterolateral thoracotomy, 3 median sternotomy) 25 VATS (5 converted to open)	86%	77%	81%	74%	14 years (median)
Lococo et al. 2012 [26]	Multicentric	50	50 malignant	2 wedge resection 2 lobectomy 4 pneumonectomy 13 tumor excision 12 extended resection 3 explorative	NA	81.1%	66.9%	72.1%	60.5%	66.2 months (mean)
Tapias et al. 2013 [21]	Monocentric	59	NA	43 wedge resection 1 segmentectomy 4 lobectomy 1 bilobectomy 5 extended resection	45 open (41 thoracotomy, 2 median sternotomy, 2 thoracoabdominal access) 14 VATS	92.4%	83.4%	87.2	72.1	12.9 years (median)

Table 2. *Cont.*

Authors and Year	Monocentric/ Multi-centric	Number of Patients	Benign/Malignant	Type of Surgery	Open/Minimally Invasive	5 Year OS	10 Year OS	5 Year DFS	10 Year DFS	Follow-Up
Franzen et al. 2014 [89]	Monocentric	42	24 benign 18 malignant	75.6% wedge resection 17.1% lobectomy 7.3% pneumonectomy	22 open 20 VATS	NA	NA	84%	67%	39 months
Ghanim et. al. 2017 [94]	Multicentric	125	104 benign 21 malignant	NA	91 open 32 VATS 2 RATS	NA	NA	77%	67%	NA
Diebold et al. 2017 [23]	Multicentric	78	NA	82% wedge resection 12% lobectomy 6% pneumonectomy	39 open (thoracotomy) 39 VATS	NA	172 months (median OS)	NA	93%	NA
Tan et al. 2018 [70]	Monocentric	82	70 benign 12 malignant	39 wedge resection 7 lobectomy 1 pneumonectomy 35 extended resection	60 open (thoracotomy) 22 VATS	NA	76% (malignant)	NA	53% (malignant)	56 months (median)
Bellini et al. 2019 [25]	Multicentric	107	79 benign 28 malignant	16 tumor resection 79 wedge resection 12 extended resection	81 open (70 thoracotomy, 1 median sternotomy, 1 sternolaparotomy, 2 clamshell) 26 VATS	NA	NA	NA	81%	7 years (median)
Zhou et al. 2020 [63]	Monocentric	70	58 benign 12 malignant	30 wedge resection 6 lobectomy 2 pneumonectomy 29 parietal pleura excision 3 extended resection	43 open (37 thoracotomy, 6 median sternotomy) 27 VATS (3 conversion)	100% (benign)	100% (benign) For malignant patients: median OS 83 months	100% (benign)	100% (benign)	91 months (mean)

Table 3. Main results of recent studies reporting recurrence features after surgery. The results of local and distant recurrence could be higher than the total number in cases where patients experienced both local and distant recurrence. NA: Not Available.

Authors and Year	Number of Patients	Number of Recurrence (Total)	Number of Recurrence (Benign)	Number of Recurrence (Malignant)	Number of Local Recurrence	Number of Distant Recurrence	Treatment of Recurrence
Magdeleinat et al. 2002 [30]	60	3	0	3	2	1	2 Surgery 1 Unknown
Cardillo et al. 2009 [71]	110	8	4	4	NA	NA	3 Surgery 4 Surgery and radiotherapy 1 Surgery and chemotherapy
Lahon et al. 2012 [27]	156	15	1.3%	19%	10	5	9 Surgery 1 Chemoradiotherapy 3 Radiotherapy 1 Chemotherapy 1 Unknown
Lococo et al. 2012 [26]	50	15	-	15	6	9	4 Surgery 11 Chemotherapy
Tapias et al. 2013 [21]	59	8	NA	NA	8	0	8 Surgery
Franzen et al. 2014 [89]	84	8	2	6	3	5	4 Surgery 4 Unknown
Ghanim et al. 2017 [94]	125	14	6	8	11	3	9 Surgery 5 Chemotherapy or radiotherapy
Tan et al. 2018 [70]	82	4	0	4	4	0	3 Surgery 1 Conservative treatment
Bellini et al. 2019 [25]	107	12	5	7	7	7	4 Surgery 4 Surgery and radiotherapy 1 Surgery and chemoradiotherapy 1 Chemotherapy 2 Unknown

8. Conclusions

Pleural SFT a relatively rare type of tumor that arises from the pleura. Thoracic surgeons should consider pSFT in their differential diagnosis in case of masses with variable dimensions that have high contact with the pleura. Despite the absence of prospective studies and standardized guidelines for their treatment, surgery should be always considered and should be planned and carried out with the aim of radical surgery, which might require extended resections. Postoperative surveillance is also of paramount importance as recurrence is possible, especially in tumors with more aggressive features, and surgery should be considered also in cases of local recurrence, if technically feasible.

Prospective studies and a more standardized classification are needed in the future to better stratify the treatment, risk of recurrence, and follow-up of these patients.

Author Contributions: Conceptualization, P.B. and F.M.; methodology, P.B. and F.M.; validation, all authors; investigation, all authors; resources, all authors; data curation, P.B. and F.M.; writing—original draft preparation, P.B., F.M., G.Q. and M.L.P.; writing—review and editing, all authors; visualization, all authors; supervision, P.B. and F.M.; project administration, P.B. All authors have read and agreed to the published version of the manuscript.

Funding: This research received no external funding.

Data Availability Statement: The data presented in this study are available in this article.

Conflicts of Interest: The authors declare no conflict of interest.

References

1. De Pinieux, G.; Karanian, M.; Le Loarer, F.; Le Guellec, S.; Chabaud, S.; Terrier, P.; Bouvier, C.; Batistella, M.; Neuville, A.; Robin, Y.M.; et al. Nationwide incidence of sarcomas and connective tissue tumors of intermediate malignancy over four years using an expert pathology review network. *PLoS ONE* **2021**, *16*, e0246958. [CrossRef]
2. Demicco, E.G.; Fritchie, K.J.; Han, A. Solitary fibrous tumour. In *Who Classification of Tumours Soft Tissue and Bone Tumours*, 5th ed.; IARC Press: Lyon, French, 2020.
3. Sung, S.H.; Chang, J.W.; Kim, J.; Lee, K.S.; Han, J.; Park, S.I. Solitary fibrous tumors of the pleura: Surgical outcome and clinical course. *Ann. Thorac. Surg.* **2005**, *79*, 303–307. [CrossRef]
4. England, D.M.; Hochholzer, L.; McCarthy, M.J. Localized benign and malignant fibrous tumors of the pleura. A clinicopathologic review of 223 cases. *Am. J. Surg. Pathol.* **1989**, *13*, 640–658. [CrossRef] [PubMed]
5. Gold, J.S.; Antonescu, C.R.; Hajdu, C.; Ferrone, C.R.; Hussain, M.; Lewis, J.J.; Brennan, M.F.; Coit, D.G. Clinicopathologic correlates of solitary fibrous tumors. *Cancer* **2002**, *94*, 1057–1068. [CrossRef]
6. Demicco, E.G.; Wagner, M.J.; Maki, R.G.; Gupta, V.; Iofin, I.; Lazar, A.J.; Wang, W.L. Risk assessment in solitary fibrous tumors: Validation and refinement of a risk stratification model. *Mod. Pathol.* **2017**, *30*, 1433–1442. [CrossRef]
7. Schweizer, L.; Koelsche, C.; Sahm, F.; Piro, R.M.; Capper, D.; Reuss, D.E.; Pusch, S.; Habel, A.; Meyer, J.; Göck, T.; et al. Meningeal hemangiopericytoma and solitary fibrous tumors carry the nab2-stat6 fusion and can be diagnosed by nuclear expression of stat6 protein. *Acta Neuropathol.* **2013**, *125*, 651–658. [CrossRef] [PubMed]
8. Robinson, D.R.; Wu, Y.M.; Kalyana-Sundaram, S.; Cao, X.; Lonigro, R.J.; Sung, Y.S.; Chen, C.L.; Zhang, L.; Wang, R.; Su, F.; et al. Identification of recurrent nab2-stat6 gene fusions in solitary fibrous tumor by integrative sequencing. *Nat. Genet.* **2013**, *45*, 180–185. [CrossRef]
9. Chmielecki, J.; Crago, A.M.; Rosenberg, M.; O'Connor, R.; Walker, S.R.; Ambrogio, L.; Auclair, D.; McKenna, A.; Heinrich, M.C.; Frank, D.A.; et al. Whole-exome sequencing identifies a recurrent nab2-stat6 fusion in solitary fibrous tumors. *Nat. Genet.* **2013**, *45*, 131–132. [CrossRef] [PubMed]
10. Barthelmeß, S.; Geddert, H.; Boltze, C.; Moskalev, E.A.; Bieg, M.; Sirbu, H.; Brors, B.; Wiemann, S.; Hartmann, A.; Agaimy, A.; et al. Solitary fibrous tumors/hemangiopericytomas with different variants of the nab2-stat6 gene fusion are characterized by specific histomorphology and distinct clinicopathological features. *Am. J. Pathol.* **2014**, *184*, 1209–1218. [CrossRef]
11. Huang, S.C.; Li, C.F.; Kao, Y.C.; Chuang, I.C.; Tai, H.C.; Tsai, J.W.; Yu, S.C.; Huang, H.Y.; Lan, J.; Yen, S.L.; et al. The clinicopathological significance of nab2-stat6 gene fusions in 52 cases of intrathoracic solitary fibrous tumors. *Cancer Med.* **2016**, *5*, 159–168. [CrossRef]
12. Tai, H.C.; Chuang, I.C.; Chen, T.C.; Li, C.F.; Huang, S.C.; Kao, Y.C.; Lin, P.C.; Tsai, J.W.; Lan, J.; Yu, S.C.; et al. Nab2-stat6 fusion types account for clinicopathological variations in solitary fibrous tumors. *Mod. Pathol.* **2015**, *28*, 1324–1335. [CrossRef] [PubMed]
13. Bahrami, A.; Lee, S.; Schaefer, I.M.; Boland, J.M.; Patton, K.T.; Pounds, S.; Fletcher, C.D. Tert promoter mutations and prognosis in solitary fibrous tumor. *Mod. Pathol.* **2016**, *29*, 1511–1522. [CrossRef]

14. Akaike, K.; Kurisaki-Arakawa, A.; Hara, K.; Suehara, Y.; Takagi, T.; Mitani, K.; Kaneko, K.; Yao, T.; Saito, T. Distinct clinicopathological features of nab2-stat6 fusion gene variants in solitary fibrous tumor with emphasis on the acquisition of highly malignant potential. *Hum. Pathol.* **2015**, *46*, 347–356. [CrossRef]
15. Machado, I.; Morales, G.N.; Cruz, J.; Lavernia, J.; Giner, F.; Navarro, S.; Ferrandez, A.; Llombart-Bosch, A. Solitary fibrous tumor: A case series identifying pathological adverse factors-implications for risk stratification and classification. *Virchows Arch.* **2020**, *476*, 597–607. [CrossRef]
16. Demicco, E.G.; Wani, K.; Ingram, D.; Wagner, M.; Maki, R.G.; Rizzo, A.; Meeker, A.; Lazar, A.J.; Wang, W.L. Tert promoter mutations in solitary fibrous tumour. *Histopathology* **2018**, *73*, 843–851. [CrossRef]
17. De Perrot, M.; Fischer, S.; Brundler, M.A.; Sekine, Y.; Keshavjee, S. Solitary fibrous tumors of the pleura. *Ann. Thorac. Surg.* **2002**, *74*, 285–293. [CrossRef]
18. Ricciardi, S.; Giovanniello, D.; Carbone, L.; Carleo, F.; Di Martino, M.; Jaus, M.O.; Mantovani, S.; Treggiari, S.; Tornese, A.; Cardillo, G. Malignant solitary fibrous tumours of the pleura are not all the same: Analysis of long-term outcomes and evaluation of risk stratification models in a large single-centre series. *J. Clin. Med.* **2023**, *12*, 966. [CrossRef]
19. Reisenauer, J.S.; Mneimneh, W.; Jenkins, S.; Mansfield, A.S.; Aubry, M.C.; Fritchie, K.J.; Allen, M.S.; Blackmon, S.H.; Cassivi, S.D.; Nichols, F.C.; et al. Comparison of risk stratification models to predict recurrence and survival in pleuropulmonary solitary fibrous tumor. *J. Thorac. Oncol.* **2018**, *13*, 1349–1362. [CrossRef] [PubMed]
20. Schirosi, L.; Lantuejoul, S.; Cavazza, A.; Murer, B.; Yves Brichon, P.; Migaldi, M.; Sartori, G.; Sgambato, A.; Rossi, G. Pleuropulmonary solitary fibrous tumors: A clinicopathologic, immunohistochemical, and molecular study of 88 cases confirming the prognostic value of de perrot staging system and p53 expression, and evaluating the role of c-kit, braf, pdgfrs (alpha/beta), c-met, and egfr. *Am. J. Surg. Pathol.* **2008**, *32*, 1627–1642. [PubMed]
21. Tapias, L.F.; Mino-Kenudson, M.; Lee, H.; Wright, C.; Gaissert, H.A.; Wain, J.C.; Mathisen, D.J.; Lanuti, M. Risk factor analysis for the recurrence of resected solitary fibrous tumours of the pleura: A 33-year experience and proposal for a scoring system. *Eur. J. Cardiothorac. Surg.* **2013**, *44*, 111–117. [CrossRef]
22. Tapias, L.F.; Mercier, O.; Ghigna, M.R.; Lahon, B.; Lee, H.; Mathisen, D.J.; Dartevelle, P.; Lanuti, M. Validation of a scoring system to predict recurrence of resected solitary fibrous tumors of the pleura. *Chest* **2015**, *147*, 216–223. [CrossRef] [PubMed]
23. Diebold, M.; Soltermann, A.; Hottinger, S.; Haile, S.R.; Bubendorf, L.; Komminoth, P.; Jochum, W.; Grobholz, R.; Theegarten, D.; Berezowska, S.; et al. Prognostic value of mib-1 proliferation index in solitary fibrous tumors of the pleura implemented in a new score—A multicenter study. *Respir. Res.* **2017**, *18*, 210. [CrossRef]
24. Demicco, E.G.; Park, M.S.; Araujo, D.M.; Fox, P.S.; Bassett, R.L.; Pollock, R.E.; Lazar, A.J.; Wang, W.L. Solitary fibrous tumor: A clinicopathological study of 110 cases and proposed risk assessment model. *Mod. Pathol.* **2012**, *25*, 1298–1306. [CrossRef]
25. Bellini, A.; Marulli, G.; Breda, C.; Ferrigno, P.; Terzi, S.; Lomangino, I.; Lo Giudice, F.; Brombin, C.; Laurino, L.; Pezzuto, F.; et al. Predictors of behaviour in solitary fibrous tumours of the pleura surgically resected: Analysis of 107 patients. *J. Surg. Oncol.* **2019**, *120*, 761–767. [CrossRef]
26. Lococo, F.; Cesario, A.; Cardillo, G.; Filosso, P.; Galetta, D.; Carbone, L.; Oliaro, A.; Spaggiari, L.; Cusumano, G.; Margaritora, S.; et al. Malignant solitary fibrous tumors of the pleura: Retrospective review of a multicenter series. *J. Thorac. Oncol.* **2012**, *7*, 1698–1706. [CrossRef] [PubMed]
27. Lahon, B.; Mercier, O.; Fadel, E.; Ghigna, M.R.; Petkova, B.; Mussot, S.; Fabre, D.; Le Chevalier, T.; Dartevelle, P. Solitary fibrous tumor of the pleura: Outcomes of 157 complete resections in a single center. *Ann. Thorac. Surg.* **2012**, *94*, 394–400. [CrossRef]
28. Cardillo, G.; Facciolo, F.; Cavazzana, A.O.; Capece, G.; Gasparri, R.; Martelli, M. Localized (solitary) fibrous tumors of the pleura: An analysis of 55 patients. *Ann. Thorac. Surg.* **2000**, *70*, 1808–1812. [CrossRef]
29. Rena, O.; Filosso, P.L.; Papalia, E.; Molinatti, M.; Di Marzio, P.; Maggi, G.; Oliaro, A. Solitary fibrous tumour of the pleura: Surgical treatment. *Eur. J. Cardiothorac. Surg.* **2001**, *19*, 185–189. [CrossRef]
30. Magdeleinat, P.; Alifano, M.; Petino, A.; Le Rochais, J.P.; Dulmet, E.; Galateau, F.; Icard, P.; Regnard, J.F. Solitary fibrous tumors of the pleura: Clinical characteristics, surgical treatment and outcome. *Eur. J. Cardiothorac. Surg.* **2002**, *21*, 1087–1093. [CrossRef] [PubMed]
31. Castaldo, V.; Domenici, D.; Biscosi, M.V.; Ubiali, P.; Miranda, C.; Zanette, G.; Mazzon, C.; Tonizzo, M. Doege-potter syndrome; a case of solitary fibrous pleura tumor associated with severe hypoglycemia: A case report in internal medicine. *Endocr. Metab. Immune Disord. Drug Targets*, 2023; *online ahead of print*.
32. Haddoub, S.; Gnetti, L.; Montanari, A.; Di Ruvo, R.; Carbognani, P.; Maccanelli, F.; Magotti, M.G.; Calderini, M.C.; Cioni, F.; Tardio, S.M. A case of solitary fibrous pleura tumor associated with severe hypoglycemia: The doege-potter's syndrome. *Acta Biomed.* **2016**, *87*, 314–317. [PubMed]
33. Jang, J.G.; Chung, J.H.; Hong, K.S.; Ahn, J.H.; Lee, J.Y.; Jo, J.H.; Lee, D.W.; Shin, K.C.; Lee, K.H.; Kim, M.J.; et al. A case of solitary fibrous pleura tumor associated with severe hypoglycemia: Doege-potter syndrome. *Tuberc. Respir. Dis.* **2015**, *78*, 120–124. [CrossRef]
34. Tay, C.K.; Teoh, H.L.; Su, S. A common problem in the elderly with an uncommon cause: Hypoglycaemia secondary to the doege-potter syndrome. *BMJ Case Rep.* **2015**, *2015*, bcr2014207995. [CrossRef] [PubMed]
35. Meng, W.; Zhu, H.H.; Li, H.; Wang, G.; Wei, D.; Feng, X. Solitary fibrous tumors of the pleura with doege-potter syndrome: A case report and three-decade review of the literature. *BMC Res. Notes* **2014**, *7*, 515. [CrossRef] [PubMed]

36. Cardillo, G.; Lococo, F.; Carleo, F.; Martelli, M. Solitary fibrous tumors of the pleura. *Curr. Opin. Pulm. Med.* **2012**, *18*, 339–346. [CrossRef]

37. Lee, K.S.; Im, J.G.; Choe, K.O.; Kim, C.J.; Lee, B.H. Ct findings in benign fibrous mesothelioma of the pleura: Pathologic correlation in nine patients. *AJR Am. J. Roentgenol.* **1992**, *158*, 983–986. [CrossRef] [PubMed]

38. Norton, S.A.; Clark, S.C.; Sheehan, A.L.; Ibrahim, N.B.; Jeyasingham, K. Solitary fibrous tumour of the diaphragm. *J. Cardiovasc. Surg.* **1997**, *38*, 685–686.

39. Versluis, P.J.; Lamers, R.J. Localized pleural fibroma: Radiological features. *Eur. J. Radiol.* **1994**, *18*, 124–125. [CrossRef]

40. Desser, T.S.; Stark, P. Pictorial essay: Solitary fibrous tumor of the pleura. *J. Thorac. Imaging* **1998**, *13*, 27–35. [CrossRef]

41. Cardinale, L.; Ardissone, F.; Garetto, I.; Marci, V.; Volpicelli, G.; Solitro, F.; Fava, C. Imaging of benign solitary fibrous tumor of the pleura: A pictorial essay. *Rare Tumors* **2010**, *2*, e1. [CrossRef]

42. Saifuddin, A.; Da Costa, P.; Chalmers, A.G.; Carey, B.M.; Robertson, R.J. Primary malignant localized fibrous tumours of the pleura: Clinical, radiological and pathological features. *Clin. Radiol.* **1992**, *45*, 13–17. [CrossRef]

43. Harrison-Phipps, K.M.; Nichols, F.C.; Schleck, C.D.; Deschamps, C.; Cassivi, S.D.; Schipper, P.H.; Allen, M.S.; Wigle, D.A.; Pairolero, P.C. Solitary fibrous tumors of the pleura: Results of surgical treatment and long-term prognosis. *J. Thorac. Cardiovasc. Surg.* **2009**, *138*, 19–25. [CrossRef]

44. Ajouz, H.; Sohail, A.H.; Hashmi, H.; Martinez Aguilar, M.; Daoui, S.; Tembelis, M.; Aziz, M.; Zohourian, T.; Brathwaite, C.E.M.; Cerfolio, R.J. Surgical considerations in the resection of solitary fibrous tumors of the pleura. *J. Cardiothorac. Surg.* **2023**, *18*, 79. [CrossRef] [PubMed]

45. Zhang, A.; Zhang, H.; Zhou, X.; Li, Z.; Li, N. Solitary fibrous tumors of the pleura shown on 18f-fdg and 68ga-dota-fapi-04 pet/ct. *Clin. Nucl. Med.* **2021**, *46*, e534–e537. [CrossRef]

46. Zhao, L.; Wang, H.; Shi, J. (18)f-fdg pet/ct characteristics of solitary fibrous tumour of the pleura: Single institution experience. *Ann. Nucl. Med.* **2022**, *36*, 429–438. [CrossRef]

47. Yeom, Y.K.; Kim, M.Y.; Lee, H.J.; Kim, S.S. Solitary fibrous tumors of the pleura of the thorax: Ct and fdg pet characteristics in a tertiary referral center. *Medicine* **2015**, *94*, e1548. [CrossRef]

48. Lococo, F.; Rufini, V.; Filice, A.; Paci, M.; Rindi, G. 68ga-dotatoc pet/ct in pleural solitary fibrous tumors. *Clin. Nucl. Med.* **2021**, *46*, e336–e338. [CrossRef]

49. Yao, K.; Zhu, L.; Wang, L.; Xia, R.; Yang, J.; Hu, W.; Yu, Z. Resection of giant malignant solitary fibrous pleural tumor after interventional embolization: A case report and literature review. *J. Cardiothorac. Surg.* **2022**, *17*, 134. [CrossRef]

50. Weiss, B.; Horton, D.A. Preoperative embolization of a massive solitary fibrous tumor of the pleura. *Ann. Thorac. Surg.* **2002**, *73*, 983–985. [CrossRef] [PubMed]

51. Perrotta, F.; Cerqua, F.S.; Cammarata, A.; Izzo, A.; Bergaminelli, C.; Curcio, C.; Guarino, C.; Grella, E.; Forzano, I.; Cennamo, A.; et al. Integrated therapeutic approach to giant solitary fibrous tumor of the pleura: Report of a case and review of the literature. *Open Med. (Wars)* **2016**, *11*, 220–225. [CrossRef] [PubMed]

52. Guo, J.; Chu, X.; Sun, Y.E.; Zhang, L.; Zhou, N. Giant solitary fibrous tumor of the pleura: An analysis of five patients. *World J. Surg.* **2010**, *34*, 2553–2557. [CrossRef]

53. Kaul, P.; Kay, S.; Gaines, P.; Suvarna, S.K.; Hopkinson, D.N.; Rocco, G. Giant pleural fibroma with an abdominal vascular supply mimicking a pulmonary sequestration. *Ann. Thorac. Surg.* **2003**, *76*, 935–937. [CrossRef]

54. Addagatla, K.; Mamtani, R.; Babkowski, R.; Ebright, M.I. Solitary fibrous tumor of the pleura with abdominal aortic blood supply. *Ann. Thorac. Surg.* **2017**, *103*, e415–e417. [CrossRef]

55. Fan, F.; Zhou, H.; Zeng, Q.; Liu, Y. Computed tomography manifestations of a malignant solitary fibrous tumour of the pleura with distinct blood supply from celiac trunk. *Eur. J. Cardiothorac. Surg.* **2014**, *45*, 1108–1110. [CrossRef] [PubMed]

56. Song, J.Y.; Kim, K.H.; Kuh, J.H.; Kim, T.Y.; Kim, J.H. Two-stage surgical treatment of a giant solitary fibrous tumor occupying the thoracic cavity. *Korean J. Thorac. Cardiovasc. Surg.* **2018**, *51*, 415–418. [CrossRef] [PubMed]

57. Van Houdt, W.J.; Westerveld, C.M.; Vrijenhoek, J.E.; van Gorp, J.; van Coevorden, F.; Verhoef, C.; van Dalen, T. Prognosis of solitary fibrous tumors: A multicenter study. *Ann. Surg. Oncol.* **2013**, *20*, 4090–4095. [CrossRef]

58. Nomori, H.; Horio, H.; Fuyuno, G.; Morinaga, S. Contacting metastasis of a fibrous tumor of the pleura. *Eur. J. Cardiothorac. Surg.* **1997**, *12*, 928–930. [CrossRef]

59. Yao, M.J.; Ding, L.; Atay, S.M.; Toubat, O.; Ebner, P.; David, E.A.; McFadden, P.M.; Kim, A.W. A modern reaffirmation of surgery as the optimal treatment for solitary fibrous tumors of the pleura. *Ann. Thorac. Surg.* **2019**, *107*, 941–946. [CrossRef] [PubMed]

60. Petrella, F.; Rizzo, S.; Solli, P.; Borri, A.; Casiraghi, M.; Tessitore, A.; Galetta, D.; Gasparri, R.; Veronesi, G.; Pardolesi, A.; et al. Giant solitary fibrous tumor of the pleura requiring left pneumonectomy. *Thorac. Cancer* **2014**, *5*, 108–110. [CrossRef]

61. Fattahi Masuom, S.H.; Bagheri, R.; Sadrizadeh, A.; Nouri Dalouee, M.; Taherian, A.; Rajaie, Z. Outcome of surgery in patients with solitary fibrous tumors of the pleura. *Asian Cardiovasc. Thorac. Ann.* **2016**, *24*, 18–22. [CrossRef]

62. Cox, J.; Leesley, H.; DeAnda, A.; Uran, D.P.; Lick, S. Resection of a giant thoracic solitary fibrous tumor treated with preoperative arterial coiling followed by a double-level thoracotomy. *J. Surg. Case Rep.* **2023**, *2023*, rjad008. [CrossRef]

63. Zhou, C.; Li, W.; Shao, J.; Zhao, J. Thoracic solitary fibrous tumors: An analysis of 70 patients who underwent surgical resection in a single institution. *J. Cancer Res. Clin. Oncol.* **2020**, *146*, 1245–1252. [CrossRef]

64. Altinok, T.; Topcu, S.; Tastepe, A.I.; Yazici, U.; Cetin, G. Localized fibrous tumors of the pleura: Clinical and surgical evaluation. *Ann. Thorac. Surg.* **2003**, *76*, 892–895. [CrossRef] [PubMed]

65. Amore, D.; Rispoli, M.; Cicalese, M.; De Rosa, I.; Rossi, G.; Corcione, A.; Buono, S.; Curcio, C. Anterior mediastinal solitary fibrous tumor resection by da vinci((r)) surgical system in obese patient. *Int. J. Surg. Case Rep.* **2017**, *38*, 163–165. [CrossRef] [PubMed]
66. Nakanishi, A.; Haruki, T.; Nakamura, H. Cowboy technique for solitary fibrous tumor of the pleura via single-port video-assisted thoracoscopic surgery. *Asian J. Endosc. Surg.* **2022**, *15*, 472–473. [CrossRef]
67. Rocco, G.; Martucci, N.; Setola, S.; Franco, R. Uniportal video-assisted thoracic resection of a solitary fibrous tumor of the pleura. *Ann. Thorac. Surg.* **2012**, *94*, 661–662. [CrossRef]
68. Bendixen, M.; Jørgensen, O.D.; Kronborg, C.; Andersen, C.; Licht, P.B. Postoperative pain and quality of life after lobectomy via video-assisted thoracoscopic surgery or anterolateral thoracotomy for early stage lung cancer: A randomised controlled trial. *Lancet Oncol.* **2016**, *17*, 836–844. [CrossRef]
69. Tamura, M.; Matsumoto, I.; Saito, D.; Yoshida, S.; Takata, M.; Takemura, H. Case report: Uniportal video-assisted thoracoscopic resection of a solitary fibrous tumor preoperatively predicted visceral pleura origin using dynamic chest radiography. *J. Cardiothorac. Surg.* **2020**, *15*, 166. [CrossRef] [PubMed]
70. Tan, F.; Wang, Y.; Gao, S.; Xue, Q.; Mu, J.; Mao, Y.; Gao, Y.; Zhao, J.; Wang, D.; Zhou, L.; et al. Solitary fibrous tumors of the pleura: A single center experience at national cancer center, china. *Thorac. Cancer* **2018**, *9*, 1763–1769. [CrossRef]
71. Cardillo, G.; Carbone, L.; Carleo, F.; Masala, N.; Graziano, P.; Bray, A.; Martelli, M. Solitary fibrous tumors of the pleura: An analysis of 110 patients treated in a single institution. *Ann. Thorac. Surg.* **2009**, *88*, 1632–1637. [CrossRef]
72. Hatooka, S.; Shigematsu, Y.; Nakanishi, M.; Yamaki, K. Subxiphoid approach for extracting a giant solitary fibrous tumour of the pleura. *Interact. Cardiovasc. Thorac. Surg.* **2017**, *25*, 834–835. [CrossRef]
73. Furukawa, N.; Hansky, B.; Niedermeyer, J.; Gummert, J.; Renner, A. A silent gigantic solitary fibrous tumor of the pleura: Case report. *J. Cardiothorac. Surg.* **2011**, *6*, 122. [CrossRef] [PubMed]
74. Filosso, P.L.; Asioli, S.; Ruffini, E.; Rovea, P.; Macri, L.; Sapino, A.; Bretti, S.; Lyberis, P.; Oliaro, A. Radical resection of a giant, invasive and symptomatic malignant solitary fibrous tumour (sft) of the pleura. *Lung Cancer* **2009**, *64*, 117–120. [CrossRef]
75. Yanagiya, M.; Matsumoto, J.; Miura, T.; Horiuchi, H. Extended thoracotomy with subcostal incision for giant solitary fibrous tumor of the diaphragm. *AME Case Rep.* **2017**, *1*, 8. [CrossRef]
76. Andrews, W.G.; Vallieres, E. Implications of adverse biological factors and management of solitary fibrous tumors of the pleura. *Thorac. Surg. Clin.* **2021**, *31*, 347–355. [CrossRef]
77. Ghanim, B.; Baier, D.; Pirker, C.; Müllauer, L.; Sinn, K.; Lang, G.; Hoetzenecker, K.; Berger, W. Trabectedin is active against two novel, patient-derived solitary fibrous pleural tumor cell lines and synergizes with ponatinib. *Cancers* **2022**, *14*, 5602. [CrossRef] [PubMed]
78. Levard, A.; Derbel, O.; Méeus, P.; Ranchère, D.; Ray-Coquard, I.; Blay, J.Y.; Cassier, P.A. Outcome of patients with advanced solitary fibrous tumors: The centre leon berard experience. *BMC Cancer* **2013**, *13*, 109. [CrossRef]
79. Schöffski, P.; Timmermans, I.; Hompes, D.; Stas, M.; Sinnaeve, F.; De Leyn, P.; Coosemans, W.; Van Raemdonck, D.; Hauben, E.; Sciot, R.; et al. Clinical presentation, natural history, and therapeutic approach in patients with solitary fibrous tumor: A retrospective analysis. *Sarcoma* **2020**, *2020*, 1385978. [CrossRef]
80. Liu, M.; Liu, B.; Dong, L.; Liu, B. Recurrent intrathoracic solitary fibrous tumor: Remarkable response to radiotherapy. *Ann. Thorac. Med.* **2014**, *9*, 245–247. [CrossRef] [PubMed]
81. Suter, M.; Gebhard, S.; Boumghar, M.; Peloponisios, N.; Genton, C.Y. Localized fibrous tumours of the pleura: 15 new cases and review of the literature. *Eur. J. Cardiothorac. Surg.* **1998**, *14*, 453–459. [CrossRef]
82. Haas, R.L.; Walraven, I.; Lecointe-Artzner, E.; Scholten, A.N.; van Houdt, W.J.; Griffin, A.M.; Ferguson, P.C.; Miah, A.B.; Zaidi, S.; DeLaney, T.F.; et al. Radiation therapy as sole management for solitary fibrous tumors (sft): A retrospective study from the global sft initiative in collaboration with the sarcoma patients euronet. *Int. J. Radiat. Oncol. Biol. Phys.* **2018**, *101*, 1226–1233. [CrossRef]
83. Saynak, M.; Bayir-Angin, G.; Kocak, Z.; Oz-Puyan, F.; Hayar, M.; Cosar-Alas, R.; Karamustafaoglu, A.; Yurut-Caloglu, V.; Caloglu, M.; Yoruk, Y. Recurrent solitary fibrous tumor of the pleura: Significant response to radiotherapy. *Med. Oncol.* **2010**, *27*, 45–48. [CrossRef] [PubMed]
84. Park, M.S.; Ravi, V.; Conley, A.; Patel, S.R.; Trent, J.C.; Lev, D.C.; Lazar, A.J.; Wang, W.L.; Benjamin, R.S.; Araujo, D.M. The role of chemotherapy in advanced solitary fibrous tumors: A retrospective analysis. *Clin. Sarcoma Res.* **2013**, *3*, 7. [CrossRef]
85. Martin-Broto, J.; Stacchiotti, S.; Lopez-Pousa, A.; Redondo, A.; Bernabeu, D.; de Alava, E.; Casali, P.G.; Italiano, A.; Gutierrez, A.; Moura, D.S.; et al. Pazopanib for treatment of advanced malignant and dedifferentiated solitary fibrous tumour: A multicentre, single-arm, phase 2 trial. *Lancet Oncol.* **2019**, *20*, 134–144. [CrossRef]
86. Maruzzo, M.; Martin-Liberal, J.; Messiou, C.; Miah, A.; Thway, K.; Alvarado, R.; Judson, I.; Benson, C. Pazopanib as first line treatment for solitary fibrous tumours: The royal marsden hospital experience. *Clin. Sarcoma Res.* **2015**, *5*, 5. [CrossRef] [PubMed]
87. Park, M.S.; Patel, S.R.; Ludwig, J.A.; Trent, J.C.; Conrad, C.A.; Lazar, A.J.; Wang, W.L.; Boonsirikamchai, P.; Choi, H.; Wang, X.; et al. Activity of temozolomide and bevacizumab in the treatment of locally advanced, recurrent, and metastatic hemangiopericytoma and malignant solitary fibrous tumor. *Cancer* **2011**, *117*, 4939–4947. [CrossRef]
88. Stacchiotti, S.; Negri, T.; Libertini, M.; Palassini, E.; Marrari, A.; De Troia, B.; Gronchi, A.; Dei Tos, A.P.; Morosi, C.; Messina, A.; et al. Sunitinib malate in solitary fibrous tumor (sft). *Ann. Oncol.* **2012**, *23*, 3171–3179. [CrossRef] [PubMed]
89. Franzen, D.; Diebold, M.; Soltermann, A.; Schneiter, D.; Kestenholz, P.; Stahel, R.; Weder, W.; Kohler, M. Determinants of outcome of solitary fibrous tumors of the pleura: An observational cohort study. *BMC Pulm. Med.* **2014**, *14*, 138. [CrossRef]

90. Kovacs, T.; Waxman, J. Recurrence of a malignant solitary fibrous tumor of the pleura 17 years after primary tumor resection—A case report. *Respir. Med. Case Rep.* **2019**, *28*, 100895. [CrossRef]

91. Von Mehren, M.; Kane, J.M.; Agulnik, M.; Bui, M.M.; Carr-Ascher, J.; Choy, E.; Connelly, M.; Dry, S.; Ganjoo, K.N.; Gonzalez, R.J.; et al. Soft Tissue Sarcoma, Version 2.2022, NCCN Clinical Practice Guidelines in Oncology. *J. Natl. Compr. Cancer Netw.* **2022**, *207*, 815–833. [CrossRef]

92. Liu, W.L.; Wu, W.; Hong, Q.C.; Lv, K. Recurrence rates of surgically resected solitary fibrous tumours of the pleura: A systematic review and meta-analysis. *Interact. Cardiovasc. Thorac. Surg.* **2021**, *32*, 882–888. [CrossRef]

93. Van Leeuwen, R.J.H.; Brunner, S.; Pojda, J.; Diebold, J.; Kestenholz, P.; Minervini, F. Intrapulmonary solitary fibrous tumor with adenofibromatous pattern with features of pleomorphic high grade sarcoma-a case report and an overview of the differential diagnosis. *Quant. Imaging Med. Surg.* **2021**, *11*, 472–478. [CrossRef] [PubMed]

94. Ghanim, B.; Hess, S.; Bertoglio, P.; Celik, A.; Bas, A.; Oberndorfer, F.; Melfi, F.; Mussi, A.; Klepetko, W.; Pirker, C.; et al. Intrathoracic solitary fibrous tumor—An international multicenter study on clinical outcome and novel circulating biomarkers. *Sci. Rep.* **2017**, *7*, 12557. [CrossRef] [PubMed]

 *cancers*

Article

Reduction of Tumor Growth with RNA-Targeting Treatment of the NAB2–STAT6 Fusion Transcript in Solitary Fibrous Tumor Models

Yi Li [1,2,†], John T. Nguyen [1,2,†], Manasvini Ammanamanchi [1], Zikun Zhou [1,2], Elijah F. Harbut [3], Jose L. Mondaza-Hernandez [4,5], Clark A. Meyer [1], David S. Moura [4], Javier Martin-Broto [4,5,6], Heather N. Hayenga [1,*] and Leonidas Bleris [1,2,7,*]

1. Department of Bioengineering, University of Texas at Dallas, Richardson, TX 75080, USA; yxl121030@utdallas.edu (Y.L.); john.nguyen12@utdallas.edu (J.T.N.); manasvini30@gmail.com (M.A.); zikun.zhou@utdallas.edu (Z.Z.); clark.meyer@utdallas.edu (C.A.M.)
2. Center for Systems Biology, University of Texas at Dallas, Richardson, TX 75080, USA
3. Department of Chemistry and Biochemistry, University of Texas at Dallas, Richardson, TX 75080, USA; elijah.harbut@utdallas.edu
4. Health Research Institute Fundacion Jimenez Diaz, Universidad Autonoma de Madrid (IIS/FJD-UAM), 28049 Madrid, Spain; jmondaza@atbsarc.org (J.L.M.-H.); dmoura@atbsarc.org (D.S.M.); jmartin@atbsarc.org (J.M.-B.)
5. University Hospital General de Villalba, 28400 Madrid, Spain
6. Medical Oncology Department, University Hospital Fundación Jimenez Diaz, 28040 Madrid, Spain
7. Department of Biological Sciences, University of Texas at Dallas, Richardson, TX 75080, USA
* Correspondence: heather.hayenga@utdallas.edu (H.N.H.); bleris@utdallas.edu (L.B.); Tel.: +972-883-5785 (L.B.)
† These authors contributed equally to this work.

 check for updates

Citation: Li, Y.; Nguyen, J.T.; Ammanamanchi, M.; Zhou, Z.; Harbut, E.F.; Mondaza-Hernandez, J.L.; Meyer, C.A.; Moura, D.S.; Martin-Broto, J.; Hayenga, H.N.; et al. Reduction of Tumor Growth with RNA-Targeting Treatment of the NAB2–STAT6 Fusion Transcript in Solitary Fibrous Tumor Models. *Cancers* **2023**, *15*, 3127. https://doi.org/10.3390/cancers15123127

Academic Editor: Bahil Ghanim

Received: 2 May 2023
Revised: 5 June 2023
Accepted: 7 June 2023
Published: 9 June 2023

Simple Summary: Solitary fibrous tumor (SFT) is a soft-tissue sarcoma occurring in adults and infants. This nonhereditary cancer is the result of an environmental intrachromosomal gene fusion between NAB2 and STAT6 on chromosome 12. Either surgery or radiation is the first line of treatment for this cancer; however for many, this becomes challenging, as the cancer can travel to inoperable areas or reoccur in locations already irradiated. Currently there is no approved chemotherapy regimen for SFTs. Anti-angiogenic drugs developed to treat other cancers have been used on SFTs with limited success. Therefore, there is a need for systemic therapy. In this study, we showed that RNA-targeting technologies (antisense oligonucleotides and CRISPR/CasRx) can be used to specifically suppress the expressions of NAB2–STAT6 fusion transcripts, but not wild type STAT6, and reduce cell proliferation and tumor growth. These results provide the foundation for a potentially novel therapeutical strategy for SFTs.

Abstract: Solitary fibrous tumor (SFT) is a rare soft-tissue sarcoma. This nonhereditary cancer is the result of an environmental intrachromosomal gene fusion between NAB2 and STAT6 on chromosome 12, which fuses the activation domain of STAT6 with the repression domain of NAB2. Currently there is not an approved chemotherapy regimen for SFTs. The best response on available pharmaceuticals is a partial response or stable disease for several months. The purpose of this study is to investigate the potential of RNA-based therapies for the treatment of SFTs. Specifically, in vitro SFT cell models were engineered to harbor the characteristic NAB2–STAT6 fusion using the CRISPR/SpCas9 system. Cell migration as well as multiple cancer-related signaling pathways were increased in the engineered cells as compared to the fusion-absent parent cells. The SFT cell models were then used for evaluating the targeting efficacies of NAB2–STAT6 fusion-specific antisense oligonucleotides (ASOs) and CRISPR/CasRx systems. Our results showed that fusion specific ASO treatments caused a 58% reduction in expression of fusion transcripts and a 22% reduction in cell proliferation after 72 h in vitro. Similarly, the AAV2-mediated CRISPR/CasRx system led to a 59% reduction in fusion transcript expressions in vitro, and a 55% reduction in xenograft growth after 29 days ex vivo.

Keywords: solitary fibrous tumor (SFT); CRISPR; antisense oligonucleotides (ASOs); RfxCas13d (CasRx)

1. Introduction

Solitary fibrous tumor (SFT) is a rare soft-tissue tumor of mesenchymal origin that accounts for less than 2% of all soft tissue sarcomas. In the latest WHO classification published in April 2020, SFTs are subdivided into three categories: benign (locally aggressive), NOS (rarely metastasizing), and malignant [1–3]. Traditionally, these tumors present a unique diagnostic challenge due to their gross and histologic features overlapping with many other soft-tissue tumors. A breakthrough occurred in 2013, when all SFTs were found to have a version of hallmark intrachromosomal gene fusion between NAB2 and STAT6 on chromosome 12 [4,5]. Research suggests that NAB2–STAT6 is the oncogenic driver [4,6,7]. Normally, early growth response-1 (EGR-1) activates NAB2, and NAB2 in turn represses oncogenic EGR-1 target genes [4]. However, in the case of the NAB2–STAT6 fusion, the activation domain of STAT6 replaces the repression domain of NAB2. As a result, instead of repressing EGR-1 target genes, the fusion protein activates them [8]. Despite these advances in the classification and understanding of the molecular mechanisms of SFTs, no SFT-specific systemic therapeutic options are yet available [9]. Anti-angiogenic drugs developed to treat other cancers have been used on SFTs with limited success [10]. None of the chemotherapies enables complete remission, with the best response being a partial response or stable disease for several months.

ASOs (antisense oligonucleotides) are synthetic single-stranded oligonucleotides or oligonucleotide analogs (typically 15–25 bp in length) which can bind to RNAs via Watson–Crick base pairing. Both protein-coding RNAs (messenger RNAs, mRNAs), noncoding RNAs such as long noncoding RNAs (lncRNAs), and microRNAs can be targeted via distinct mechanisms, including the RNase H-dependent and RNase H-independent pathways [11–13]. Various chemical modification methods, such as phosphorothioate and 2′-O-methoxyethyl (MOE) modification, are commonly used to enhance the binding affinity of ASOs to target RNAs, increase metabolic stability, and decrease possible adverse effects. To date, eight ASOs have been approved by the FDA for disorders including Duchenne muscular dystrophy, spinal muscular atrophy, and cytomegalovirus (CMV) retinitis [14–22].

CRISPR (clustered regularly interspaced palindromic repeats) technology has revolutionized the field of genetic engineering [23–26]. Various CRISPR-mediated DNA editing technologies, such as the type II CRISPR/SpCas9 system or type V CRISPR-Cas12 (Cpf1) system, have been extensively studied and already moved into clinical trials [27–29]. In parallel, efforts have been made to discover and establish diverse Cas effectors which can precisely manipulate RNA molecules. More recently, the Type VI CRISPR/RfxCas13d (CasRx) system, which is derived from *Ruminococcus flavefaciens* XPD3002 and possesses dual RNase activities, was found to efficiently process target RNAs in mammalian cells with fewer non-specific targeting effects compared to other RNA-editing technologies such as shRNAs (short hairpin RNA) [30–32].

We note that all NAB2–STAT6 fusion types in SFTs create unique fusion RNA transcripts which are distinct from wild type NAB2 or STAT6 transcripts. Therefore, this study investigated both ASO- and CasRx-based RNA targeting technologies to specifically suppress the expression level of NAB2–STAT6 fusion transcripts, which we hypothesized would exert anti-tumor benefits for SFTs. We report suppression of RNA fusion transcripts and associated reduction in growth rates in vitro and in xenograft models, pointing to a potentially viable therapeutic strategy for SFTs.

2. Materials and Methods

2.1. Mammalian Cell Culture

All cells investigated herein, that is HCT116 (American Type Culture Collection, Manassas, VA, USA catalog number: CCL-247), NS-poly, NS-11, NS-17, and NS-23, were maintained at 37 °C, 100% humidity and 5% CO_2. The cells were grown in Dulbecco's modified Eagle's medium (DMEM media, Invitrogen, Waltham, MA, USA, catalog number: 11965-1181) supplemented with 10% fetal bovine serum (FBS, Invitrogen, catalog number: 26140), 0.1 mM MEM non-essential amino acids (Invitrogen, catalog number: 11140-050), and 0.045 units/mL of Penicillin and 0.045 units/mL of Streptomycin (Penicillin-Streptomycin liquid, Invitrogen, catalog number: 15140). To pass the cells, the adherent culture was first washed with PBS (Dulbecco's Phosphate Buffered Saline, Mediatech, Manassas, VA, USA, catalog number: 21-030-CM), then trypsinized with Trypsin-EDTA (0.25% Trypsin with EDTAX4Na, Invitrogen, catalog number: 25200) and finally diluted in fresh medium. Additionally, for maintaining NS-poly cells, hygromycin (200 µg/mL, Thermo Fisher Scientific, Waltham, MA, USA, catalog number: 10687010) was included in the complete growth medium.

2.2. Long-Range Genomic PCR and RT (Reverse Transcription)-PCR

For long-range genomic PCR, total genomic DNAs were harvested using DNeasy Blood & Tissue kit (Qiagen, Germantown, MD, USA, catalog number: 69504), and long-range PCR reactions were performed using Q5 High-Fidelity 2× Master Mix (New England Biolabs, Ipswich, WA, USA, catalog number: M0492). An amount of 100 ng genomic DNAs were used as the template, and the PCR conditions were: first, 1 cycle of 98 °C for 30 s, followed by 40 cycles of 98 °C for 10 s, 66 °C for 30 s, and 72 °C for 2 min. The PCR products were subjected to 1% agarose gel electrophoresis and the DNA bands of interest were purified using QIAquick Gel Extraction Kit (Qiagen, catalog number: 28706) and subjected to direct Sanger sequencing (Genewiz) and analyzed using FinchTV (Geospiza). Specifically, for NAB2exon6-STAT6exon17 fusion allele, the forward primer (5′-GGTCATGTCCAAGGCTGACGCCGCCCCCTG-3′) is located within exon 6 of NAB2, and reverse primer (5′-GTAGCTGGGACATAACCCCTGCCATCCTTACC-3′) is located within exon 17 of STAT6. For wild type NAB2 allele, the forward primer (5′-GGTCATGTCCAAGGCTGACGCCGCCCCCTG-3′) is in exon 6 of NAB2 gene; and the reverse primer (5′-CCTCCCCTCCCTGGCTGTGCGTAGCTCTGT-3′) is in intron 6 of NAB2 gene.

For RT-PCR, total RNAs were harvested using RNeasy Mini kit (Qiagen, catalog number: 74106) and cDNAs were made using QuantiTect Reverse Transcription kit (500 ng RNA, Qiagen, catalog number: 205311). The cDNAs were then subjected to PCR reactions using Q5 High-Fidelity 2× Master Mix and the PCR conditions were: first, 1 cycle of 98 °C for 30 s, followed by 40 cycles of 98 °C for 10 s, 63 °C for 30 s, and 72 °C for 1 min. Specifically, for the NAB2exon6-STAT6exon17 fusion transcript, the forward primer (5′-CCTGTCTGGGGAGAGTCTGGATG-3′) is in exon 5 of NAB2 gene, and the reverse primer (5′-GGGGGGATGGAGTGAGAGTGTG-3′) is in exon 20 of STAT6 gene. The PCR products were subjected to 1% agarose gel electrophoresis and the DNA bands of interest were purified using QIAquick Gel Extraction Kit and subjected to direct Sanger sequencing (Genewiz) and analyzed using FinchTV (Geospiza).

2.3. Western Blot

To prepare whole cell lysates, cell pellets were washed with ice-cold PBS and then resuspended in RIPA buffer (Cell Signaling, catalog number: 9806) with protease/phosphatase inhibitors (Cell Signaling, catalog number: 5872). The resuspended cells were homogenized 10 times using a 1 mL syringe with 30 Gauge needles (VWR, catalog number: 328411), followed by 30 min incubation on ice. The cell suspensions were then centrifuged at 16,000× g rpm for 10 min and the supernatants were collected, and the protein concentrations were determined using Pierce BCA Protein Assay kit (Thermo Fisher Scientific, catalog number: 23227). For Western blot, 30 µg protein were used with anti-STAT6 C-terminus primary antibody (1:1000, Abcam, Cambridge, UK, catalog number: ab32520) and mouse anti-rabbit horseradish peroxidase-conjugated secondary antibody (1:5000, Santa Cruz Biotechnology, catalog number: sc-2357). The signal was visualized using the

ChemiDoc XRS+ imaging system (BIO-RAD, Hercules, CA, USA, catalog number: 1708265) The reversible Ponceau staining was used as the alternative loading control [33,34].

2.4. Cell Proliferation Assay

80,000 HCT116 or NS-poly cells were seeded onto a 12-well plate (Greiner Bio-One, Monroe, NC, USA, catalog number: 665180) without hygromycin. At 24, 48 and 72 h, cells were trypsinized with 250 µL 0.25% trypsin-EDTA at 37 °C for 5 min. Trypsin-EDTA was then neutralized by adding 750 µL of complete medium. The cell suspension was then counted using a hemacytometer (Sigma-Aldrich, St. Louis, MO, USA, catalog number: Z359629). All experiments were performed in triplicates.

2.5. Wound Healing Assay

400,000 HCT116 or NS-poly cells were seeded onto a 6-well plate (Greiner Bio-One, catalog number: 657160) without hygromycin. After 48 h, the growth media was removed, and the confluent cells were scratched from top to bottom of the well using a 20 µL pipet tip. The cells were then gently washed with PBS twice to remove any cell debris. Finally, 2 mL of DMEM media supplemented with 0.1 mM MEM non-essential amino acids, 0.045 units/mL of Penicillin, and 0.045 units/mL of Streptomycin was added. The brightfield images (magnification 10×) were then captured every 2 h (up to 48 h) using an Olympus IX81 microscope (Tokyo, Japan). Data collection and processing were performed in the software package Slidebook 5.0.

2.6. RNA Sequencing (RNA-Seq)

For RNA-seq, total RNAs were harvested using RNeasy Mini kit and sample purities were evaluated using OD260/OD280 (1.8–2.2). The Genewiz Standard RNA-Seq service was employed, which requires 2 µg of total RNAs in 10 µL nuclease-free water. The assays were performed on an Illumina HiSeq platform (2 × 150 bp configuration, single index) and the outputs contained ~50 million reads per sample. All samples had mean quality scores larger than 30, and the summary sequencing statistics were presented in Supplementary Table S1. For data analysis, the human reference genome (UCSC hg38, https://genome-idx.s3.amazonaws.com/hisat/hg38_tran.tar.gz (accessed on 1 September 2021)) was used, and a pipeline consisting of HISAT2, StringTie and DESeq2 was employed to identify differentially expressed genes between the wild type HCT116 and NS-poly cells.

2.7. ASO Treatment

All ASOs were dissolved in PBS. ASO treatment was performed either with transfection reagents (RNAiMAX, Thermo Fisher Scientific, catalog number: 13778975) or without transfection reagents (free delivery by directly adding to the growth media). Cells were seeded in complete growth medium. After 18 h, ASOs were delivered using RNAiMAX following the manufacturer's recommendation or added directly into the growth medium. After 48 h, the cells were harvested for further analysis.

2.8. Real-Time RT-PCR (Reverse Transcription-PCR)

For real-time RT-PCR assays, total RNAs were extracted using RNeasy Mini Kit (Qiagen, catalog number: 74106). First-strand cDNAs were synthesized using QuantiTect Reverse Transcription kit (500 ng RNA, Qiagen, catalog number: 205311). Next, quantitative PCR was performed using the KAPA SYBR FAST universal qPCR Kit (Kapa Biosystems, Wilmington, MA, USA, catalog number: KK4601), with GAPDH as the internal control. The forward primer (P25) for GAPDH was: 5′-AATCCCATCACCATCTTCCA-3′ and the reverse primer (P26) for GAPDH was: 5′-TGGACTCCACGACGTACTCA-3′. Quantitative analysis was performed using the $2^{-\Delta\Delta Ct}$ method. Fold-change values were reported as means with standard deviations.

2.9. Recombinant AAV2 Viral Vector Production

For primary AAV2 viral vector production, the AAV-2 Helper Free System (Agilent, catalog number: 240071) was used. Briefly, AAV-293 cells were seeded at 70–80% confluency. The cells were then transfected with pAAV-RC, pHelper and pAAV expression plasmid using JetPRIME (Polyplus Transfection, catalog number: 101000046). The cells were harvested after 72 h and subjected to four rounds of freeze/thaw cycles using a dry ice-ethanol bath and a 37 °C water bath. The cells were then centrifuged at 16,000 g for 10 min at room temperature, and the supernatant was transferred to new Eppendorf tubes stored at -80 °C. The physical titers of primary AAV2 viral vectors were determined by qPCR AAV Titer Kit (Applied Biological Materials, Richmond, BC, Canada, catalog number: G931).

2.10. Fluorescence Microscopy

Microscopy was performed 48 h post-transfection. The live cells were grown on 12-well plates (Greiner Bio-One) in the complete medium. Cells were imaged using an Olympus IX81 microscope (Tokyo, Japan) in a precision-controlled environmental chamber. The images were captured using a Hamamatsu ORCA-03 cooled monochrome digital camera (Hamamatsu, Japan). The filter sets (Chroma Technology, Bellows Falls, VT, USA) are as follows: ET500/20× (excitation) and ET535/30 m (emission) for Yellow Fluorescent Protein (YFP). Data collection and processing was performed in the software package Slidebook 5.0. All images within a given experimental set were collected with the same exposure times and underwent identical processing.

2.11. Flow Cytometry

A total of 48 h post-transfection, cells from each well of the 12-well plates were trypsinized with 250 μL 0.25% trypsin-EDTA at 37 °C for 5 min. Trypsin-EDTA was then neutralized by adding 750 μL of complete medium. The cell suspension was centrifuged at 1000 rotations per minute for 5 min, and after removal of supernatants, the cell pellets were resuspended in 0.5 mL phosphate-buffered saline buffer. The cells were analyzed on a BD Biosciences LSRFortessa flow analyzer (San Jose, CA, USA). YFP was measured with a 488 nm laser, a 535 nm emission filter and a 545/35 band-pass filter. For data analysis, 100,000 events were collected. A forward scatter/side scatter gate was generated using an un-transfected negative sample and applied to all cell samples. All experiments were performed in triplicates.

2.12. Mouse Xenograft

Foxn1^{nu} athymic nude mice were purchased from The Jackson Laboratory (Bar Harbor, ME, USA, catalog number: 002019) and maintained in a pathogen-free facility, in accordance with Protocol #21-05 approved by IACUC of University of Texas at Dallas. For mouse xenograft, 5 million NS-poly cells were resuspended in 100 μL of PBS and injected subcutaneously into the right flank region of 4–6-week-old female Foxn1^{nu} athymic mice under anesthesia using isoflurane (Covetrus, Portland, ME, USA, catalog number: 11695-6777-2). The tumor size was measured using a digital caliper and the tumor volume was calculated as $(L \times W \times W)/2$, where L was tumor length and W the tumor width. Mice were monitored daily for general health (signs of dehydration, cachexia, weight loss, and dyspnea) and euthanized when tumors reached 2 cm^3 in volume or mouse body weight decreased by more than 20%.

2.13. H and E (Hematoxylin and Eosin) Staining

The paraffin-embedded tissue slides (10 μm) were first deparaffinized by heating at 60 °C for 10 min. The slides were then re-hydrated and staining using hematoxylin 560 (Leica, Wetzlar, Germany, catalog number: 3801570). The slides were then counterstained using Eosin Y 515 (Leica, catalog number: 3801615). After dehydration, one drop of mounting medium (Abcam, catalog number: ab64230) and a glass cover were added to each slide, and slides were then observed using a Leica DMi1 Microscopy.

2.14. CT Scan

Each mouse underwent micro-CT (OI/CT, MILabs, Utrecht, The Netherlands) imaging using an accurate, ultra-focus image scan at a step angle of 0.250 degrees at 1 projection per step and a binning size of 1. The micro-CT tube settings were set at a voltage of 50 kV, a current of 0.21 mA, and an exposure time of 75 ms. Images were converted to DICOMs using vendor software OsiriX (version 12.0.0).

3. Results

3.1. Generation of NAB2exon6–STAT6exon17 Gene Fusion Cell Models Using Genome Editing

A recent breakthrough in understanding SFTs was the identification of recurrent NAB2–STAT6 gene fusions in almost all SFT tissue samples [4,5]. Specifically, the breakpoints within NAB2 and STAT6 genes, which are adjacent on chromosome 12q13, induce the inversion of DNA fragments and subsequently the expression of NAB2–STAT6 transactivators.

Thus far, seven distinct NAB2–STAT6 gene fusion types (NAB2exon2–STAT6exon5, NAB2exon4–STAT6exon2, NAB2exon4–STAT6exon4, NAB2exon5–STAT6exon16, NAB2exon6–STAT6exon16, NAB2exon6–STAT6exon17, and NAB2exon7–STAT6exon2) have been discovered to commonly account for pathologic variation and tumor aggressiveness in SFTs. Herein all exons and introns of the human STAT6 gene were numbered in accordance with the latest NCBI reference transcript (NM_001178078.2).

We first aimed to use the CRISPR/spCas9 system to generate an HCT116-based NAB2exon6–STAT6exon17 fusion cell model. We noted that for this fusion type, the inverted sequence was approximately 5 kb (Figure 1a, from NAB2 intron 6 to STAT6 intron 16), and thus within the size limit for a SpCas9-mediated knock-in strategy (typically less than 10 kb) [35–40]. Here, the human colorectal cell line HCT116 was employed due to its high transfection and editing efficiency, as compared to SFTs' (debated) cell of origin of mesenchymal cells [41]. Our engineered cell lines did indeed recapitulate pathogenic tumor aspects, allowed us to profile the impact of the fusion to their transcriptome, and assess the efficacy of RNA-targeting methodologies in vitro and in vivo.

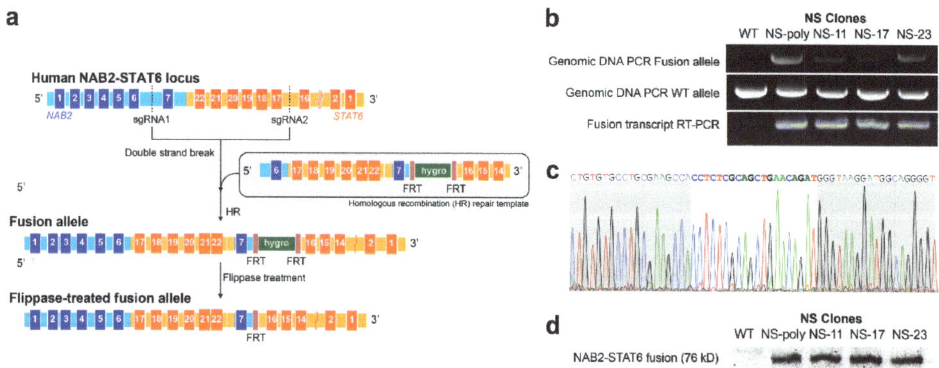

Figure 1. Preparation of HCT116-based NAB2exon6–STAT6exon17 fusion stable cell lines using the CRISPR/SpCas9 system. (**a**) Schematic illustration of the CRISPR/SpCas9 homologous recombination process to create NAB2exon6–STAT6exon17 fusion type in HCT116 cells. (**b**) Genotyping of NAB2exon6–STAT6exon17 fusion stable cells. Genomic DNA PCR-and RNA RT-PCR assays confirmed that the fusion stable cells were heterozygous. (**c**) Sanger sequencing result for RT-PCR amplicons confirmed the fusion type. (**d**) Western blot confirmed the expression of NAB2exon6–STAT6exon17 fusion proteins in the fusion stable cells.

Two sgRNAs were designed to target intron 6 of NAB2 gene (5′-CAGAAATTCCAGC-GCAACCGAGG-3′, PAM underlined) and intron 16 of STAT6 gene (5′-GGAGGAAGTGGG-TGACAGGAAGG-3′, PAM underlined), respectively. Next, a homologous recombination

(HR) repair template, which contains the inverted sequence of original NAB2 intron 6-STAT6 intron 16, was designed (Figure 1a). In addition, a hygromycin resistance gene cassette, which was flanked by two FRT (flippase recognition target) sites, was placed after the exon 7 of NAB2. HCT116 cells were transiently transfected with PCMV–SpCas9, the two sgRNA constructs, along with the HR repair template, and after 48 h, treated with 200 μg/mL hygromycin for two weeks. The established polyclonal stable cell line (named as NS-poly) was further transfected with a flippase-expressing construct to remove the hygromycin resistance gene cassette. Finally, the transfected cells were sorted into single cells using flow cytometry cell sorter, and three monoclonal stable cell lines were established (named as NS-11, NS-17, and NS-23).

To determine the genotype of our stable cells, we first harvested genomic DNAs using DNeasy Blood & Tissue kit and performed genotyping PCR reactions using primers P1 and P2 (Supplementary Table S2) for the NAB2exon6–STAT6exon17 fusion allele. As shown in Figure 1b, the expected amplicon (1194 bp) was observed in NS-poly, NS-11, NS-17, and NS-23, but not in HCT116 wild type cells, indicating that all four stable cell lines contained the fusion allele. Furthermore, we subjected the genomic DNAs to PCR reactions using primers P3 and P4, and the expected amplicon (1020 bp) was observed in all stable cells as well as in the wild-type sample. Taken together, these results showed that all four stable cell lines are heterozygous.

We further confirmed successful integration of the NAB2–STAT6 gene fusion at the RNA level. We extracted total RNAs from the stable cells using RNeasy kit and subjected the RNA samples to RT-PCR for the fusion transcript (441 bp, forward primer P5: 5′-CCTGTCTGGGGAGAGTCTGGATG-3′, exon 5 of NAB2 gene; reverse primer P6: 5′-GGGGGGATGGAGTGAGAGTGTG-3′, exon 20 of STAT6 gene). Like our genomic PCR results, all four stable cell lines yielded the expected band, which was absent in the original HCT116 cells (Figure 1b). Subsequent Sanger sequencing further confirmed the fusion type as NAB2exon6–STAT6exon17 (Figure 1c, breakpoint adjacent region between NAB2 exon 6: 5′-CCTCTCGCAG-3′ and STAT6 exon 17: 5′-CTGAACAGAT-3′ highlighted in white). Lastly, we prepared whole cell lysates using RIPA buffer, which were subsequently subjected to Western blot using a STAT6 C-terminus-targeting antibody. As shown in Figure 1d and Supplementary Figure S1, all four fusion stable cells expressed the NAB2exon6–STAT6exon17 fusion protein (expected size: 76 kD), which was not observed in the HCT116 wild type cells.

For in vitro characterization of NS-poly and its parental HCT116 cell line (HCT116), we first performed cell proliferation assay using a hemacytometer, which showed no significant difference of cell growth rates between the two cell lines (Supplementary Figure S2). Next, the wound healing assays were used to measure the cell migratory potentials (Supplementary Figure S3a, for wild type HCT116 at 0 and 48 h), which showed that the NS-poly cells had a higher motility rate compared to HCT116 cells (Supplementary Figure S3b). As an example (Supplementary Table S3), 24 h after scratching, the NS-poly cells closed the wound by 283.5 pixels, compared to 202.8 pixels for wild type HCT116 cells.

3.2. RNA-Sequencing (RNA-Seq) Analysis of NAB2exon6–STAT6exon17 Gene Fusion Cell Models

NAB2–STAT6 fusions have been shown to function as transcriptional activators and upregulate the expression of cancer-promoting EGR1 target genes including FGFR1 and NTRK1, as well as IGF genes in SFT patient samples [4,5]. To systemically characterize the pathway and network perturbations induced by NAB2–STAT6 fusions, we subjected our HCT116-based NAB2–STAT6 fusion stable cells to RNA-sequencing (RNA-seq). Briefly, total RNAs were harvested from HCT116 wild type (3 replicates) as well as NS-11, NS-17, and NS-23 cells (one sample for each monoclonal cell line), and subsequently RNA-seq assays were performed on an Illumina HiSeq platform (Genewiz). Next, an analysis pipeline consisting of HISAT2, StringTie and DESeq2 was used to identify differentially expressed genes between the wild type and fusion cells [42]. As shown in Figure 2a (adjusted *p*-values < 0.01 using FDR/Benjamini–Hochberg multiple testing correction) and Figure 2b, the expression of NAB2–STAT6 fusion induced extensive changes at the transcriptional levels (Supplementary Table S4, 198 differentially expressed genes using relatively stringent filtering conditions: adjusted *p*-values < 0.01, and | log$_2$(fold-change) | > 3). Subsequently, we performed signaling pathway analysis using the PANTHER classification system and identified candidate genes which contained both MSTRG numbers and corresponding gene names (131 genes, Supplementary Table S5) [43]. As shown in Figure 2c and Supplementary Table S6, multiple cancer-related signaling pathways have been identified, including the FGF signaling pathway, VEGF signaling pathway, EGF receptor signaling pathway, and Ras pathway. Additionally, the identified angiogenesis pathway is consistent with previous reports of the interfacing between STAT6 and neoangiogenesis [44]. We emphasize that analyses of these perturbed genes and signaling pathways could provide crucial insights into the development of SFT cancers, as well as allow rational design of targeted therapeutic options. Specifically, we observed that a particular glycosyltransferase family member, MGAT5 (Mannoside Acetylglucosaminyltransferase 5), was overexpressed in all three fusion monoclonal cell lines, but absent in wild type control replicates (Figure 2b), which implied that protein glycosylation may play a role in SFT pathology, if similar results can be observed in additional engineered or primary SFT cell models.

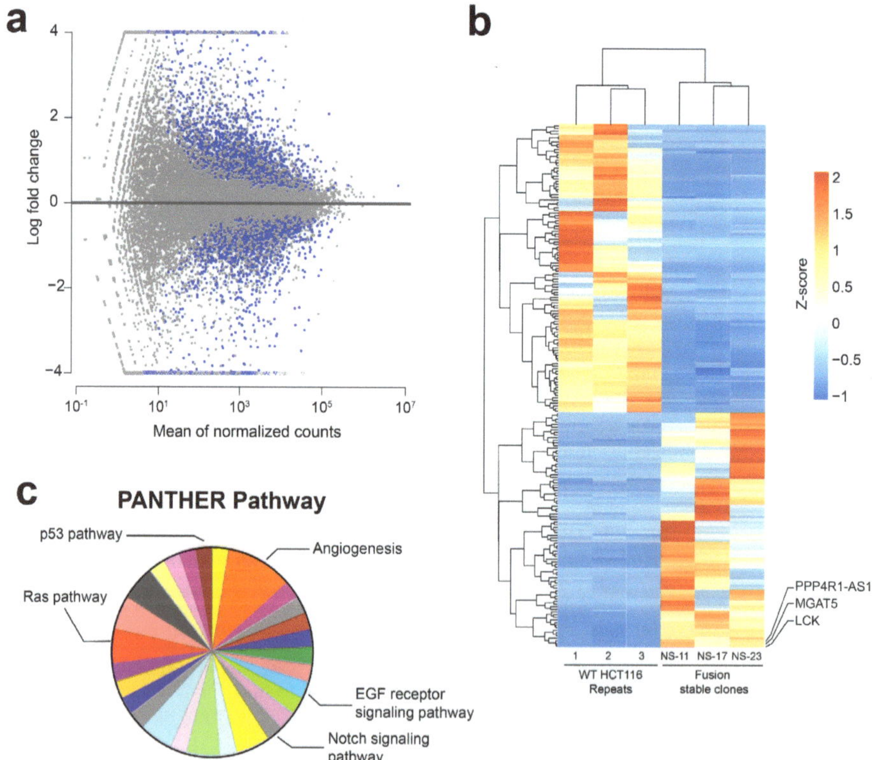

Figure 2. Transcriptome analysis of monoclonal NAB2exon6–STAT6exon17 fusion cells with their control HCT116 cells. (**a**) MA-plot analysis of monoclonal fusion cells vs. control HCT116 cells displaying differentially expressed (blue) genes (adjusted *p*-values < 0.01). (**b**) Heatmap showing differentially expressed genes between fusion stable cells and control HCT116. Expressions of MGAT5 were upregulated in three monoclonal fusion cells. (**c**) PANTHER analysis showed that multiple cancer-related signaling pathways were affected in monoclonal fusion cells.

3.3. In Vitro Targeting of NAB2–STAT6 Fusion Transcripts Using Fusion-Specific Antisense Oligonucleotides (ASOs)

Unlike the noncancerous cells, in SFT cancer cells the fusion of NAB2 and STAT6 transcripts create novel sequences at the junction site (e.g., 5′-CCTCTCGCAG | CTGAACAGAT-3′ for NAB2exon6–STAT6exon17 fusion type, | denotes the breakpoint), which could serve as the basis for fusion-specific RNA targeting. Accordingly, for NS-poly cells, we designed three fusion-specific ASOs with 2′-O-methoxyethyl modifications (fusion2, fusion4, and fusion6, Supplementary Table S7). Subsequently, we tested the in vitro targeting efficacies against the NAB2–STAT6 fusion transcripts (NAB2–STAT6 fusion-specific primers, P21 and P22, Supplementary Table S2) using both liposome-based (Lipofectamine RNAiMAX, Invitrogen) and free delivery methods.

A control ASO (Integrated DNA Technologies, 5′-C*G*T*T*A*A*T*C*G*C*G*T*A*T*A*-A*T*A*C*G*-3′, * denotes phosphorothioate modification), designed not to target any human RNA transcripts, was also included. As shown in Figure 3a, using the RNAiMAX delivery (1 μM final concentration due to possible non-specific cytotoxicity effects of ASOs at higher concentrations [45–47]), all three candidate ASOs suppressed the expression of NAB2–STAT6 fusion transcript, with Fusion6 ASO yielded the highest efficacy (58% suppression, *p*-value < 0.05). In contrast, none of the three ASOs induced any significant suppression using the free delivery method (Figure 3b). These results implied that our cell line model, which is colonic epithelial HCT116-based, may not be compatible with the free uptake mechanism.

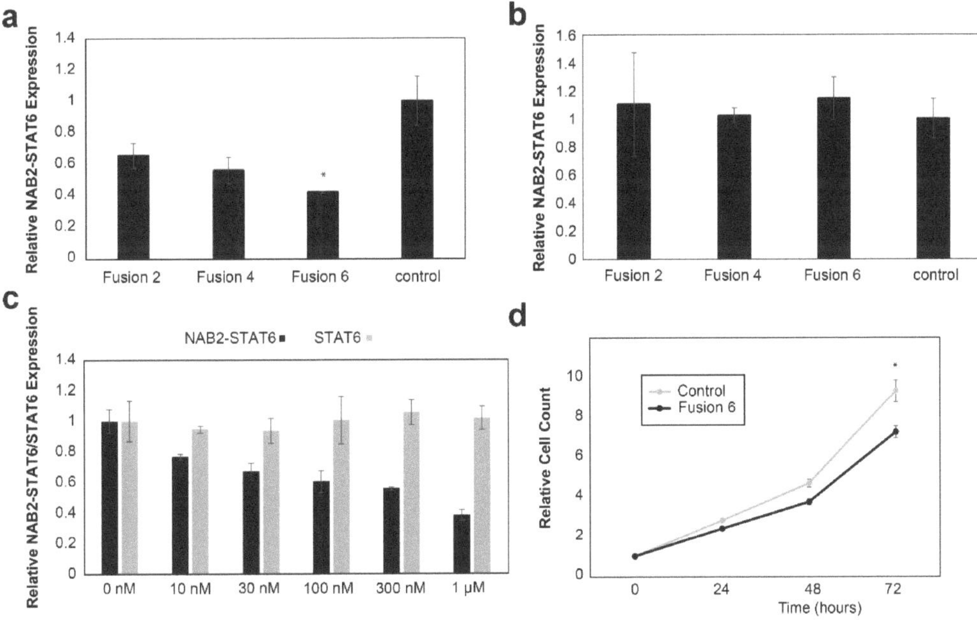

Figure 3. In vitro targeting of NAB2–STAT6 fusion transcripts using fusion-specific ASOs. (**a**) In vitro targeting of NAB2–STAT6 fusion transcripts by ASOs using RNAiMAX-mediated delivery. Fusion6 ASO suppressed the expression of NAB2–STAT6 transcript by 58%. (**b**) In vitro targeting of NAB2–STAT6 fusion transcripts by ASOs using gymnotic delivery. (**c**) Fusion6 ASO efficiently and specifically suppressed the expression of NAB2–STAT6 transcript. (**d**) Fusion6 ASO suppressed the proliferation of NS-poly cells compared to the negative control ASO. (* indicates *p*-value < 0.05 using two-tailed *t*-test).

Next, NS-poly cells were treated with Fusion6 ASO using RNAiMAX at different dosages (0 nm, 10 nm, 30 nm, 100 nm, 300 nm, and 1 µM). As shown in Figure 3c, while Fusion6 ASO efficiently suppressed the expression of NAB2–STAT6 transcript (primers P21 and P22, IC50: 356.6 nM), no suppression of wild type STAT6 transcript (wild type STAT6-specific primers P23 and P24, Supplementary Table S2) was observed. Lastly, we transfected NS-poly cells with either Fusion6 or the control ASO at 1 µM concentration. As shown in Figure 3d, Fusion6 ASO significantly suppressed the proliferation of NS-poly cells after 72 h (22.1% suppression, *p*-value < 0.05). Our results indicated that Fusion6 ASO-based RNA targeting can affect both RNA expression and cell proliferation rates in NS-poly cells.

3.4. In Vitro Targeting of NAB2–STAT6 Fusion Transcripts Using AAV2-Mediated Fusion-Specific CRISPR/CasRx

The *Rfx*Cas13d (CasRx)-based RNA editing has been reported to be highly efficient in mammalian cells [30–32]. Additionally, the presence of a PAM (protospacer adjacent motif) is not required for Cas binding. Thus, for our NAB2 exon6–STAT6 exon17 fusion type, a 30–nt fusion mRNA junction-targeting pre-gRNA sequence was designed (5'-TACCCATCTGTTCAG|CTGCGAGAGGTGGCT-3', | denotes the break point), and subsequently cloned into a CMV-CasRx-U6-sgRNA.NAB2STAT6 construct (Figure 4a). We first evaluated its editing efficacies using a YFP (Yellow Fluorescent Protein) reporter construct, which contained the corresponding target site after the start codon ATG. As shown in Figure 4b,c, upon delivery into HCT116 cells, this CasRx–pre-sgRNA complex potently suppressed the expression of YFP by 97% (*p*-value < 0.001), compared to a negative control (NC) sgRNA which targets the human *AAVS1* locus (5'-ACCCAGAACCAGAGCCACATTAACCGGCCC-3').

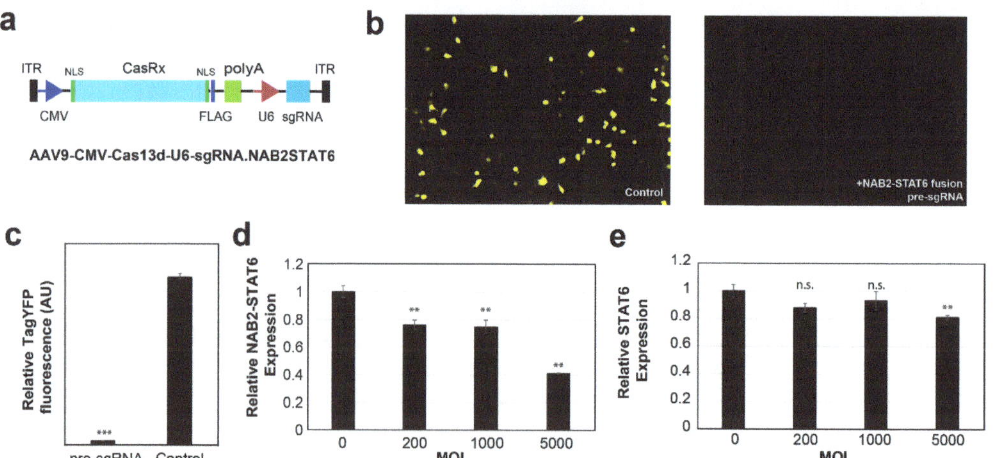

Figure 4. In vitro targeting of NAB2–STAT6 fusion transcripts using fusion-specific CRISPR–CasRx system. (**a**) Schematic illustration of the AAV2-based CRISPR/CasRx construct. (**b**) Fluorescence microscopy and (**c**) flow cytometry assay showed that the CRISPR/CasRx efficiently suppressed YFP expression, whose transcript contains a CasRx target site after the start codon (ATG). (**d**) Quantitative RT-PCR showed that AAV2-based NAB2–STAT6 fusion targeting viral vectors suppressed the expression of NAB2–STAT6 fusion transcripts in NS-poly cells. (**e**) AAV2-based NAB2–STAT6 fusion targeting viral vectors only mildly suppressed the expression of wild type STAT6 fusion transcript in NS-poly cells at the highest tested MOI (19% at MOI 5000). Two-tailed *t*-tests were used for all statistical testing. *** indicates *p*-value < 0.001, ** indicated *p*-value < 0.01, and n.s. indicates *p*-value > 0.05.

Next, we noted that CasRx (930 aa) is the smallest discovered member in the Cas13 protein family, which makes it fully compatible with most AAV packaging systems. Thus, we prepared the NAB2 exon6-STAT6 exon17 fusion-targeting CasRx AAV2 viral vectors, and transduced NS-poly cells at various MOIs (Multiplicity of Infection). As shown in Figure 4d, the viral vectors suppressed the expression of the NAB2–STAT6 fusion transcript in a dose-dependent manner (25% suppression at MOI 1000, p-value < 0.01, and 59% suppression at MOI 5000, p-value < 0.01). In contrast, the vectors exerted no statistically significant suppression effects on the wild type STAT6 transcript at lower MOIs (Figure 4e, p-value > 0.20 at MOI 1000). It should be noted that a mild suppression of STAT6 transcript was observed at higher MOIs (19% suppression at MOI 5000, p-value < 0.01) which may indicate AAV2-associated cell toxicity [48]. Take together, these results demonstrated that our AAV2-mediated CasRx system was both effective and specific on the fusion transcript at lower MOIs, and the MOI of 1000 was used for the subsequent ex vivo targeting experiments.

3.5. Ex Vivo Targeting of NAB2–STAT6 Fusion Transcripts Using AAV2-Mediated Fusion-Specific CRISPR/CasRx System

To evaluate whether AAV2-mediated NAB2–STAT6 fusion-targeting CasRx system could exert in vivo therapeutic benefits for SFTs, we first created a subcutaneous (subQ) NS-poly xenograft model using 5 million NS-poly cells and Foxn1^{nu} athymic nude mice (The Jackson Laboratory). As an example, 5 weeks post-injection, one xenograft measured ~2.3 cm^3, which was then harvested for preparation of paraffin sections (10 μm). Subsequent Hematoxylin and eosin (H and E) staining showed both necrotic core and neovascular blood vessels within the mass (Figure 5a).

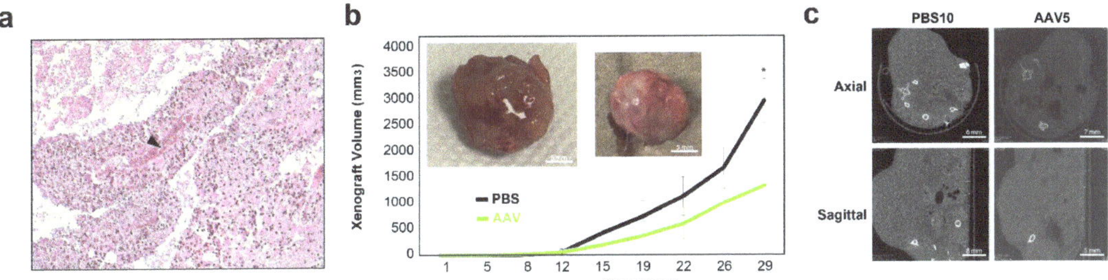

Figure 5. Ex vivo targeting of NAB2–STAT6 fusion transcripts using fusion-specific CRISPR–CasRx system. (**a**) Hematoxylin and eosin (H and E) staining of NS-polys cells-derived mouse xenograft. The arrow indicates the formation of neovascular blood vessels in the xenograft tissue. (**b**) Tumor growth curves of PBS-(black) or NAB2-STAT6 CasRx (green)-treated NS-poly cells subcutaneously injected in Foxn1^{nu} athymic nude mice. From 29 days post-implantation, NAB2–STAT6 CasRx-treated group showed slower tumor growth (* indicates p-value < 0.05 using two-tailed t-test). (inlet) Representative images of xenograft harvested from each treatment group; PBS-treated (**left**), NAB2-STAT6 CasRx-treated (**right**). (**c**) Representative high resolution CT scan images of mouse cross sections around the tumor; PBS-treated (**left**), NAB2–STAT6 CasRx-treated (**right**).

For ex vivo evaluation, NS-poly cells were transduced with NAB2–STAT6 fusion-targeting CasRx AAV2 viral vectors (NAB2-STAT6 CasRx) at MOI of 1000 in a 10 cm Petri dish. After 48 h, treated cells were harvested for xenograft and tumor sizes were measured twice a week using a digital caliper. PBS-treated NS-poly cells were used as the negative control. As shown in Figure 5b, 29 days post-implantation, tumor sizes were significantly smaller in the NAB2–STAT6 CasRx-treated mice (1340.7 mm^3 compared to 2963.4 mm^3 for PBS-treated mice, p-value < 0.05), which were corroborated by high-resolution computed tomography (CT) scans with axial and sagittal views of the tumors prior to excision (Figure 5c).

Taken together, our data demonstrated the potential therapeutic benefits of AAV2-mediated fusion-specific CRISPR/CasRx system in suppressing the SFT tumor growth.

4. Discussion

A major bottleneck to targeted therapy development is the lack of cell models of SFT. Using the CRISPR genome editing technologies, we have built several in vitro SFT cell line models (e.g., NS-poly) in HCT116 cells, partially due to their high transfection and genome editing efficiencies. Having the HCT116 cells with and without the NAB2–STAT6 gene fusion is crucial to isolating the impact of the fusion's contribution alone on the pathologic variation and tumor aggressiveness. These cell lines better resemble primary SFTs compared to cell models used in previous studies. As an example, a NAB2–STAT6 fusion cDNA construct was stably integrated into mouse fibroblast cell line NIH-3T3 [7]. The resulting cells, although partially recovering characteristics of SFTs, do not preserve important genetic information of the original NAB2–STAT6 due to the lack of sequences including endogenous NAB2 promoters, 5′-UTRs of NAB2, and 3′-UTRs of STAT6. In contrast, our CRISPR-generated cell models preserved all original NAB2–STAT6 gene fusion information.

Our results here highlight the oncogenic role of the NAB2–STAT6 in SFT. Namely, compared to the wild type HCT116 cells, the presence of the NAB2–STAT6 fusion led to an increase in cell migration. These results suggest the downstream pathways of the gene fusion may influence cell motility. RNA-sequencing results indicate cell motility and growth signaling pathways are affected by the NAB2–STAT6 fusion (e.g., RAS, EGFR, VEGF, FGF, etc.). Encouragingly, fusion specific ASO treatment reduced the fusion transcript by 58% and cell proliferation by 22% after 72 h in vitro. Similarly, AAV2-mediated CRISPR/CasRx treatments reduced the fusion transcript by 59% in vitro and tumor growth by 55% after 29 days ex vivo. Notably, the parent wild type HCT116 cells are inherently oncogenic. Therefore, the fact that an observable reduction in the growth of the HCT116 cells with the NAB2–STAT6 gene was obtained by exclusively reversing the NAB2–STAT6 expression, further emphasizes the role of the NAB2–STAT6 gene fusion in tumorigenicity.

In this study, we have designed ASOs which specifically target the NAB2–STAT6 fusion transcripts but not wild type NAB2 or STAT6 transcripts (Figure 3). It should be noted that this design strategy depends on specific NAB2–STAT6 fusion types (NAB2exon6-STAT6exon17 type in our case), which increases the specificity and reduces adverse off-target side-effects. However, the cost of a personalized approach may limit its broad clinical use. There have been shown to be at least six distinct fusion types that may account for pathologic variation and tumor aggressiveness seen in SFTs [6]. Alternatively, we note that all known NAB2–STAT6 fusions contain the C-terminal of the human STAT6 transcript, which includes the 3′-UTR (untranslated region) sequence (~1.1 kb). More importantly, previous studies have demonstrated that depletion of wild type STAT6 may elicit beneficial effects [44,49–51]. Therefore, novel ASOs could be designed to target the 3′-UTR of STAT6, which could in theory suppress the expression levels of both wild type STAT6 and all known NAB2–STAT6 fusion transcripts, especially if a SFT-tumor-specific ASO delivery method is developed.

Next, we note that for CRISPR/CasRx systems, there are discrepancies in the current literature with respect to their specificities, or the collateral degradation of non-target RNAs [52,53]. As an example, You and colleagues showed that the collateral activities of the CasRx system positively correlate with the abundance of target RNA, which could subsequently induce the cleavage of 28 s rRNA and cell toxicity. Furthermore, both CasRx-reactive antibodies and CasRx-responding T cells have been reported in healthy human donors [54]. Although not observed in our CasRx experiments, RNA editing may elicit cytotoxicity. Taken together, thorough biosafety studies of CasRx are required before being applied to clinical treatments.

Finally, we do want to highlight the limitations of our current study. Firstly, although our engineered cells provide a means to differentially isolate the effects of the fusion, our choice of cell line (HCT116) may not be ideal. SFTs are believed to originate from mesenchymal stem cells with fibroblastic differentiation, but not from colonic epithelial cells. Accordingly, although our RNA-seq assay and the subsequent PANTHER signaling pathway analysis points to the potential involvement of VEGF and EFGR signaling, their clinical significance in SFT remains unclear, as both RAS and TP53 genes are reportedly mutated in HCT116 and both genes are known to have interactions with VEGF or EFGR signaling [55,56]. Independent of the cell type of origin, primary occurrences are observed throughout the body with various presentations. For example, Bieg M. et al. noted that pleuropulmonary SFTs are less cellular and more collagenous, whereas retroperitoneum, pelvic, and meningeal SFTs have more ovoid or round cell morphologies [8]. Thus, the slight uncertainty in presentation and cell type of origin, renders the study of the fusion behavior in different cellular backgrounds still insightful. Notably, although not necessarily derived from cholinergic cells, SFTs have been reportedly present as a polyp in the colon tissue [57].

In follow up studies, we will engineer all the NAB2–STAT6 fusions in mesenchymal fibroblastic cell lines, and subsequently re-examine the targeting effects of our candidate ASOs and the CRISPR/CasRx system in these engineered cell models and SFT patient derived primary cell lines (e.g., SFT-T1 and SFT-T2 cell lines from Ghanim et al. [58]). Similarly, we will confirm our signaling pathway analysis results in these new cell models using RNA-seq assays, and upon confirmation, perform additional immunohistochemistry (IHC) or immunofluorescence (IF) assays for protein targets in such signaling (e.g., PI3K for VEGF signaling [59]) using the corresponding ex vivo tumor tissues.

Additionally, it should be emphasized that the use of ASOs remains challenging in a clinical setting, especially in terms of the delivery of the drug to the desired site (e.g., solid tumor tissues). Briefly, ASO drugs need to travel through the blood stream, pass through biological barriers, and withstand lysosomal degradation upon internalization by target cells. To address these difficulties, different chemical modification methods, such as the phosphorothioate (PS) backbone, 2′-MOE, and locked nucleic acids (LNA) have been developed to increase their stability. Additionally, various delivery vehicles, from DNA nanostructures to exosomes, have been used to increase the delivery efficiency.

Lastly, the ex vivo efficacy of our NAB2–STAT6 fusion-targeting CRISPR/CasRx system was evaluated using a xenograft model, which may not fully capture the pathological demonstrations of SFT. Thus, in future studies, we plan to reassess candidate RNA-based therapeutics (ASOs and CRISPR/CasRx) using our recently developed SFT PDX (patient-derived xenograft) model [60], and additionally explore the combinations of ASOs and CRISPR/CasRx for their potential synergistic therapeutic effects.

5. Conclusions

Our study demonstrated the potential of using RNA therapeutics (antisense oligonucleotides and the CRISPR/CasRx system) to target the pathological NAB2–STAT6 fusion transcripts in SFTs. Further investigations are needed to evaluate their in vivo efficacies and safety before being translated into clinical practice.

Supplementary Materials: The following supporting information can be downloaded at: https://www.mdpi.com/article/10.3390/cancers15123127/s1, Figure S1: Original data for Western blot; Figure S2: cell proliferation assays for wild type HCT116 and NS-poly cells; Figure S3: in vitro characterization of NS-poly cell model; Table S1: summary sequencing statistics of the RNA-seq assay; Table S2: primers used in this study; Table S3: wound healing assays using HCT116 and NS-poly cells; Table S4: Differentially expressed genes between wild type HCT116 and NAB2–STAT6 fusion cells from RNA-seq; Table S5: differentially expressed genes used for PANTHER signaling pathway analysis; Table S6: NAB2–STAT6 fusion-related signaling pathways identified by the PANTHER classification system; Table S7: NAB2–STAT6 fusion-specific ASOs with 2′-O-methoxyethyl modifications; prepDE.py: the python script used to quantify both expressed genes and transcripts; DESeq2.r: the R script used to identify differentially expressed genes and prepare for the corresponding heatmap.

Author Contributions: Conceptualization: Y.L., J.T.N., C.A.M., H.N.H. and L.B.; Formal Analysis: Y.L. and J.T.N.; Investigation: Y.L., J.T.N., Z.Z., E.F.H. and M.A.; Writing—Original Draft Preparation: Y.L., J.T.N., C.A.M., J.M.-B., D.S.M., J.L.M.-H., H.N.H. and L.B.; Supervision: H.H and L.B. All authors have read and agreed to the published version of the manuscript.

Funding: LB acknowledges funding from the US National Science Foundation (NSF) grant 2114192, a Cecil H. and Ida Green Endowment, and the University of Texas at Dallas. HH acknowledges funding from the University of Texas at Dallas Bioengineering Transform Grant and Vice President Accelerator Award.

Institutional Review Board Statement: The study was conducted in accordance with the Declaration of Helsinki, and approved by the IACUC of University of Texas at Dallas (protocol number: #21-05, date of approval: 14 June 2021).

Informed Consent Statement: Not applicable.

Data Availability Statement: Data collected for this article are available in the Supplementary Materials.

Acknowledgments: We thank the laboratory members in the Bleris and Hayenga labs for their support and discussions.

Conflicts of Interest: The authors declare no conflict of interest.

References

1. Sheehan, J.; Kondziolka, D.; Flickinger, J.; Lunsford, L.D.; Coffey, R.J.; Loeffler, J.S.; Sawaya, R.; Gutin, P.H. Radiosurgery for Treatment of Recurrent Intracranial Hemangiopericytomas. *Neurosurgery* **2002**, *51*, 905–911. [CrossRef] [PubMed]
2. Ali, H.S.M.; Endo, T.; Endo, H.; Murakami, K.; Tominaga, T. Intraspinal Dissemination of Intracranial Hemangiopericytoma: Case Report and Literature Review. *Surg. Neurol. Int.* **2016**, *7* (Suppl. S40), S1016–S1020. [CrossRef]
3. Galanis, E.; Buckner, J.C.; Scheithauer, B.W.; Kimmel, D.W.; Schomberg, P.J.; Piepgras, D.G. Management of Recurrent Meningeal Hemangiopericytoma. *Cancer* **1998**, *82*, 1915–1920. [CrossRef]
4. Robinson, D.R.; Wu, Y.M.; Kalyana-Sundaram, S.; Cao, X.; Lonigro, R.J.; Sung, Y.S.; Chen, C.L.; Zhang, L.; Wang, R.; Su, F.; et al. Identification of Recurrent NAB2-STAT6 Gene Fusions in Solitary Fibrous Tumor by Integrative Sequencing. *Nat. Genet.* **2013**, *45*, 180–185. [CrossRef] [PubMed]
5. Chmielecki, J.; Crago, A.M.; Rosenberg, M.; O'Connor, R.; Walker, S.R.; Ambrogio, L.; Auclair, D.; McKenna, A.; Heinrich, M.C.; Frank, D.A.; et al. Whole-Exome Sequencing Identifies a Recurrent NAB2-STAT6 Fusion in Solitary Fibrous Tumors. *Nat. Genet.* **2013**, *45*, 131–132. [CrossRef]
6. Guseva, N.V.; Tanas, M.R.; Stence, A.A.; Sompallae, R.; Schade, J.C.; Bossler, A.D.; Bellizzi, A.M.; Ma, D. The NAB2–STAT6 Gene Fusion in Solitary Fibrous Tumor Can Be Reliably Detected by Anchored Multiplexed PCR for Targeted next-Generation Sequencing. *Cancer Genet.* **2016**, *209*, 303–312. [CrossRef]
7. Park, Y.S.; Kim, H.S.; Kim, J.H.; Choi, S.H.; Kim, D.S.; Ryoo, Z.Y.; Kim, J.Y.; Lee, S. NAB2-STAT6 Fusion Protein Mediates Cell Proliferation and Oncogenic Progression via EGR-1 Regulation. *Biochem. Biophys. Res. Commun.* **2020**, *526*, 287–292. [CrossRef] [PubMed]
8. Bieg, M.; Moskalev, E.A.; Will, R.; Hebele, S.; Schwarzbach, M.; Schmeck, S.; Hohenberger, P.; Jakob, J.; Kasper, B.; Gaiser, T.; et al. Gene Expression in Solitary Fibrous Tumors (SFTs) Correlates with Anatomic Localization and NAB2-STAT6 Gene Fusion Variants. *Am. J. Pathol.* **2021**, *191*, 602–617. [CrossRef]
9. Gao, Q.; Liang, W.W.; Foltz, S.M.; Mutharasu, G.; Jayasinghe, R.G.; Cao, S.; Liao, W.W.; Reynolds, S.M.; Wyczalkowski, M.A.; Yao, L.; et al. Driver Fusions and Their Implications in the Development and Treatment of Human Cancers. *Cell Rep.* **2018**, *23*, 227–238.e3. [CrossRef] [PubMed]
10. de Bernardi, A.; Dufresne, A.; Mishellany, F.; Blay, J.-Y.; Ray-Coquard, I.; Brahmi, M. Novel Therapeutic Options for Solitary Fibrous Tumor: Antiangiogenic Therapy and Beyond. *Cancers* **2022**, *14*, 1064. [CrossRef] [PubMed]
11. Zogg, H.; Singh, R.; Ro, S. Current Advances in RNA Therapeutics for Human Diseases. *Int. J. Mol. Sci.* **2022**, *23*, 2736. [CrossRef]
12. Shadid, M.; Badawi, M.; Abulrob, A. Antisense Oligonucleotides: Absorption, Distribution, Metabolism, and Excretion. *Expert Opin. Drug Metab. Toxicol.* **2021**, *17*, 1281–1292. [CrossRef]
13. Gupta, A.; Andresen, J.L.; Manan, R.S.; Langer, R. Nucleic Acid Delivery for Therapeutic Applications. *Adv. Drug Deliv. Rev.* **2021**, *178*, 113834. [CrossRef]
14. Saifullah; Motohashi, N.; Tsukahara, T.; Aoki, Y. Development of Therapeutic RNA Manipulation for Muscular Dystrophy. *Front. Genome Ed.* **2022**, *4*, 863651. [CrossRef]
15. Sartorius, K.; Antwi, S.O.; Chuturgoon, A.; Roberts, L.R.; Kramvis, A. RNA Therapeutic Options to Manage Aberrant Signaling Pathways in Hepatocellular Carcinoma: Dream or Reality? *Front. Oncol.* **2022**, *12*, 891812. [CrossRef] [PubMed]
16. Aimo, A.; Castiglione, V.; Rapezzi, C.; Franzini, M.; Panichella, G.; Vergaro, G.; Gillmore, J.; Fontana, M.; Passino, C.; Emdin, M. RNA-Targeting and Gene Editing Therapies for Transthyretin Amyloidosis. *Nat. Rev. Cardiol.* **2022**, *19*, 655–667. [CrossRef]

17. Tarn, W.-Y.; Cheng, Y.; Ko, S.-H.; Huang, L.-M. Antisense Oligonucleotide-Based Therapy of Viral Infections. *Pharmaceutics* **2021**, *13*, 2015. [CrossRef] [PubMed]

18. Grabowska-Pyrzewicz, W.; Want, A.; Leszek, J.; Wojda, U. Antisense Oligonucleotides for Alzheimer's Disease Therapy: From the MRNA to MiRNA Paradigm. *EBioMedicine* **2021**, *74*, 103691. [CrossRef]

19. Edinoff, A.N.; Nguyen, L.H.; Odisho, A.S.; Maxey, B.S.; Pruitt, J.W.; Girma, B.; Cornett, E.M.; Kaye, A.M.; Kaye, A.D. The Antisense Oligonucleotide Nusinersen for Treatment of Spinal Muscular Atrophy. *Orthop. Rev.* **2021**, *13*, 24934. [CrossRef] [PubMed]

20. Wiggins, R.; Feigin, A. Emerging Therapeutics in Huntington's Disease. *Expert Opin. Emerg. Drugs* **2021**, *26*, 295–302. [CrossRef]

21. Amado, D.A.; Davidson, B.L. Gene Therapy for ALS: A Review. *Mol. Ther.* **2021**, *29*, 3345–3358. [CrossRef] [PubMed]

22. Robson, F.; Khan, K.S.; Le, T.K.; Paris, C.; Demirbag, S.; Barfuss, P.; Rocchi, P.; Ng, W.-L. Coronavirus RNA Proofreading: Molecular Basis and Therapeutic Targeting. *Mol. Cell* **2020**, *79*, 710–727. [CrossRef] [PubMed]

23. Mali, P.; Aach, J.; Stranges, P.B.; Esvelt, K.M.; Moosburner, M.; Kosuri, S.; Yang, L.; Church, G.M. CAS9 Transcriptional Activators for Target Specificity Screening and Paired Nickases for Cooperative Genome Engineering. *Nat. Biotechnol.* **2013**, *31*, 833–838. [CrossRef] [PubMed]

24. Cong, L.; Ran, F.A.; Cox, D.; Lin, S.; Barretto, R.; Habib, N.; Hsu, P.D.; Wu, X.; Jiang, W.; Marraffini, L.A.; et al. Multiplex Genome Engineering Using CRISPR/Cas Systems. *Science* **2013**, *339*, 819–823. [CrossRef]

25. Cho, S.W.; Kim, S.; Kim, J.M.; Kim, J.S. Targeted Genome Engineering in Human Cells with the Cas9 RNA-Guided Endonuclease. *Nat. Biotechnol.* **2013**, *31*, 230–232. [CrossRef]

26. Kuscu, C.; Arslan, S.; Singh, R.; Thorpe, J.; Adli, M. Genome-Wide Analysis Reveals Characteristics of off-Target Sites Bound by the Cas9 Endonuclease. *Nat. Biotechnol.* **2014**, *32*, 677–683. [CrossRef]

27. Zarei, A.; Razban, V.; Hosseini, S.E.; Tabei, S.M.B. Creating Cell and Animal Models of Human Disease by Genome Editing Using CRISPR/Cas9. *J. Gene Med.* **2019**, *21*, e3082. [CrossRef]

28. Karimian, A.; Azizian, K.; Parsian, H.; Rafieian, S.; Shafiei-Irannejad, V.; Kheyrollah, M.; Yousefi, M.; Majidinia, M.; Yousefi, B. CRISPR/Cas9 Technology as a Potent Molecular Tool for Gene Therapy. *J. Cell. Physiol.* **2019**, *234*, 12267–12277. [CrossRef]

29. Chen, M.; Mao, A.; Xu, M.; Weng, Q.; Mao, J.; Ji, J. CRISPR-Cas9 for Cancer Therapy: Opportunities and Challenges. *Cancer Lett.* **2019**, *447*, 48–55. [CrossRef]

30. Konermann, S.; Lotfy, P.; Brideau, N.J.; Oki, J.; Shokhirev, M.N.; Hsu, P.D. Transcriptome Engineering with RNA-Targeting Type VI-D CRISPR Effectors. *Cell* **2018**, *173*, 665–676.e14. [CrossRef]

31. Tong, H.; Huang, J.; Xiao, Q.; He, B.; Dong, X.; Liu, Y.; Yang, X.; Han, D.; Wang, Z.; Wang, X.; et al. High-Fidelity Cas13 Variants for Targeted RNA Degradation with Minimal Collateral Effects. *Nat. Biotechnol.* **2023**, *41*, 108–119. [CrossRef] [PubMed]

32. Wessels, H.H.; Méndez-Mancilla, A.; Guo, X.; Legut, M.; Daniloski, Z.; Sanjana, N.E. Massively Parallel Cas13 Screens Reveal Principles for Guide RNA Design. *Nat. Biotechnol.* **2020**, *38*, 722–727. [CrossRef] [PubMed]

33. Romero-Calvo, I.; Ocón, B.; Martínez-Moya, P.; Suárez, M.D.; Zarzuelo, A.; Martínez-Augustin, O.; de Medina, F.S. Reversible Ponceau Staining as a Loading Control Alternative to Actin in Western Blots. *Anal. Biochem.* **2010**, *401*, 318–320. [CrossRef] [PubMed]

34. Goasdoue, K.; Awabdy, D.; Bjorkman, S.T.; Miller, S. Standard Loading Controls Are Not Reliable for Western Blot Quantification across Brain Development or in Pathological Conditions. *Electrophoresis* **2016**, *37*, 630–634. [CrossRef] [PubMed]

35. Li, Y.; Mendiratta, S.; Ehrhardt, K.; Kashyap, N.; White, M.A.; Bleris, L. Exploiting the CRISPR/Cas9 PAM Constraint for Single-Nucleotide Resolution Interventions. *PLoS ONE* **2016**, *11*, e0144970. [CrossRef]

36. Li, Y.; Nowak, C.M.; Withers, D.; Pertsemlidis, A.; Bleris, L. CRISPR-Based Editing Reveals Edge-Specific Effects in Biological Networks. *CRISPR J.* **2018**, *1*, 286–293. [CrossRef] [PubMed]

37. Nowak, C.M.C.M.; Lawson, S.; Zerez, M.; Bleris, L. Guide RNA Engineering for Versatile Cas9 Functionality. *Nucleic Acids Res.* **2016**, *44*, 9555–9564. [CrossRef]

38. Moore, R.; Spinhirne, A.; Lai, M.J.; Preisser, S.; Li, Y.; Kang, T.; Bleris, L. CRISPR-Based Self-Cleaving Mechanism for Controllable Gene Delivery in Human Cells. *Nucleic Acids Res.* **2015**, *43*, 1297–1303. [CrossRef]

39. Quarton, T.; Kang, T.; Papakis, V.; Nguyen, K.; Nowak, C.; Li, Y.; Bleris, L. Uncoupling Gene Expression Noise along the Central Dogma Using Genome Engineered Human Cell Lines. *Nucleic Acids Res.* **2020**, *48*, 9406–9413. [CrossRef]

40. Hsieh, M.-H.; Choe, J.H.; Gadhvi, J.; Kim, Y.J.; Arguez, M.A.; Palmer, M.; Gerold, H.; Nowak, C.; Do, H.; Mazambani, S.; et al. P63 and SOX2 Dictate Glucose Reliance and Metabolic Vulnerabilities in Squamous Cell Carcinomas. *Cell Rep.* **2019**, *28*, 1860–1878.e9. [CrossRef]

41. Davanzo, B.; Emerson, R.E.; Lisy, M.; Koniaris, L.G.; Kays, J.K. Solitary Fibrous Tumor. *Transl. Gastroenterol. Hepatol.* **2018**, *3*, 94. [CrossRef] [PubMed]

42. Pertea, M.; Kim, D.; Pertea, G.M.; Leek, J.T.; Salzberg, S.L. Transcript-Level Expression Analysis of RNA-Seq Experiments with HISAT, StringTie and Ballgown. *Nat. Protoc.* **2016**, *11*, 1650–1667. [CrossRef] [PubMed]

43. Mi, H.; Ebert, D.; Muruganujan, A.; Mills, C.; Albou, L.-P.; Mushayamaha, T.; Thomas, P.D. PANTHER Version 16: A Revised Family Classification, Tree-Based Classification Tool, Enhancer Regions and Extensive API. *Nucleic Acids Res.* **2021**, *49*, D394–D403. [CrossRef] [PubMed]

44. Binnemars-Postma, K.; Bansal, R.; Storm, G.; Prakash, J. Targeting the Stat6 Pathway in Tumor-Associated Macrophages Reduces Tumor Growth and Metastatic Niche Formation in Breast Cancer. *FASEB J.* **2018**, *32*, 969–978. [CrossRef]

45. Prakash, T.P.; Yu, J.; Shen, W.; De Hoyos, C.L.; Berdeja, A.; Gaus, H.; Liang, X.-H.; Crooke, S.T.; Seth, P.P. Site-Specific Incorporation of 2′,5′-Linked Nucleic Acids Enhances Therapeutic Profile of Antisense Oligonucleotides. *ACS Med. Chem. Lett.* **2021**, *12*, 922–927. [CrossRef]
46. Shen, W.; De Hoyos, C.L.; Migawa, M.T.; Vickers, T.A.; Sun, H.; Low, A.; Bell, T.A., 3rd; Rahdar, M.; Mukhopadhyay, S.; Hart, C.E.; et al. Chemical Modification of PS-ASO Therapeutics Reduces Cellular Protein-Binding and Improves the Therapeutic Index. *Nat. Biotechnol.* **2019**, *37*, 640–650. [CrossRef]
47. Kamola, P.J.; Maratou, K.; Wilson, P.A.; Rush, K.; Mullaney, T.; McKevitt, T.; Evans, P.; Ridings, J.; Chowdhury, P.; Roulois, A.; et al. Strategies for In Vivo Screening and Mitigation of Hepatotoxicity Associated with Antisense Drugs. *Mol. Ther. Nucleic Acids* **2017**, *8*, 383–394. [CrossRef]
48. Howard, D.B.; Powers, K.; Wang, Y.; Harvey, B.K. Tropism and Toxicity of Adeno-Associated Viral Vector Serotypes 1, 2, 5, 6, 7, 8, and 9 in Rat Neurons and Glia in Vitro. *Virology* **2008**, *372*, 24–34. [CrossRef]
49. Kamerkar, S.; Leng, C.; Burenkova, O.; Jang, S.C.; McCoy, C.; Zhang, K.; Dooley, K.; Kasera, S.; Zi, T.; Sisó, S.; et al. Exosome-Mediated Genetic Reprogramming of Tumor-Associated Macrophages by ExoASO-STAT6 Leads to Potent Monotherapy Antitumor Activity. *Sci. Adv.* **2022**, *8*, eabj7002. [CrossRef]
50. Lesterhuis, W.J.; Punt, C.J.A.; Hato, S.V.; Eleveld-Trancikova, D.; Jansen, B.J.H.; Nierkens, S.; Schreibelt, G.; de Boer, A.; Van Herpen, C.M.L.; Kaanders, J.H.; et al. Platinum-Based Drugs Disrupt STAT6-Mediated Suppression of Immune Responses against Cancer in Humans and Mice. *J. Clin. Investig.* **2011**, *121*, 3100–3108. [CrossRef]
51. Haselager, M.V.; Thijssen, R.; Bax, D.; Both, D.; De Boer, F.; Mackay, S.; Dubois, J.; Mellink, C.; Kater, A.P.; Eldering, E. JAK-STAT Signaling Shapes the NF-KB Response in CLL towards Venetoclax Sensitivity or Resistance via Bcl-XL. *Mol. Oncol.* **2022**. [CrossRef]
52. Shi, P.; Murphy, M.R.; Aparicio, A.O.; Kesner, J.S.; Fang, Z.; Chen, Z.; Trehan, A.; Guo, Y.; Wu, X. Collateral Activity of the CRISPR/RfxCas13d System in Human Cells. *Commun. Biol.* **2023**, *6*, 334. [CrossRef]
53. Li, Y.; Xu, J.; Guo, X.; Li, Z.; Cao, L.; Liu, S.; Guo, Y.; Wang, G.; Luo, Y.; Zhang, Z.; et al. The Collateral Activity of RfxCas13d Can Induce Lethality in a RfxCas13d Knock-in Mouse Model. *Genome Biol.* **2023**, *24*, 20. [CrossRef] [PubMed]
54. Tang, X.-Z.E.; Tan, S.X.; Hoon, S.; Yeo, G.W. Pre-Existing Adaptive Immunity to the RNA-Editing Enzyme Cas13d in Humans. *Nat. Med.* **2022**, *28*, 1372–1376. [CrossRef] [PubMed]
55. Farhang Ghahremani, M.; Goossens, S.; Nittner, D.; Bisteau, X.; Bartunkova, S.; Zwolinska, A.; Hulpiau, P.; Haigh, K.; Haenebalcke, L.; Drogat, B.; et al. P53 Promotes VEGF Expression and Angiogenesis in the Absence of an Intact P21-Rb Pathway. *Cell Death Differ.* **2013**, *20*, 888–897. [CrossRef]
56. Okawa, T.; Michaylira, C.Z.; Kalabis, J.; Stairs, D.B.; Nakagawa, H.; Andl, C.D.; Johnstone, C.N.; Klein-Szanto, A.J.; El-Deiry, W.S.; Cukierman, E.; et al. The Functional Interplay between EGFR Overexpression, HTERT Activation, and P53 Mutation in Esophageal Epithelial Cells with Activation of Stromal Fibroblasts Induces Tumor Development, Invasion, and Differentiation. *Genes Dev.* **2007**, *21*, 2788–2803. [CrossRef]
57. Katerji, R.; Agostini-Vulaj, D. Solitary Fibrous Tumor Presenting as a Colonic Polyp: Report of a Case and Literature Review. *Hum. Pathol. Case Rep.* **2021**, *25*, 200547. [CrossRef]
58. Ghanim, B.; Baier, D.; Pirker, C.; Müllauer, L.; Sinn, K.; Lang, G.; Hoetzenecker, K.; Berger, W. Trabectedin Is Active against Two Novel, Patient-Derived Solitary Fibrous Pleural Tumor Cell Lines and Synergizes with Ponatinib. *Cancers* **2022**, *14*, 5602. [CrossRef]
59. Karar, J.; Maity, A. PI3K/AKT/MTOR Pathway in Angiogenesis. *Front. Mol. Neurosci.* **2011**, *4*, 51. [CrossRef]
60. Mondaza-Hernandez, J.L.; Moura, D.S.; Lopez-Alvarez, M.; Sanchez-Bustos, P.; Blanco-Alcaina, E.; Castilla-Ramirez, C.; Collini, P.; Merino-Garcia, J.; Zamora, J.; Carrillo-Garcia, J.; et al. ISG15 as a Prognostic Biomarker in Solitary Fibrous Tumour. *Cell. Mol. Life Sci.* **2022**, *79*, 434. [CrossRef]

Article

Value of Cellular Components and Focal Dedifferentiation to Predict the Risk of Metastasis in a Benign-Appearing Extra-Meningeal Solitary Fibrous Tumor: An Original Series from a Tertiary Sarcoma Center

Mohammad Hassani [1,2], Sungmi Jung [3], Elaheh Ghodsi [4], Leila Seddigh [5], Paul Kooner [1], Ahmed Aoude [1] and Robert Turcotte [1,*]

[1]   Division of Orthopedic Surgery, McGill University Health Centre, Montreal, QC H3G 1A4, Canada
[2]   Department of Orthopedic Surgery, Mashhad University of Medical Sciences, Mashhad 91886-17871, Iran
[3]   Department of Pathology, McGill University Health Centre, Montreal, QC H4A 3J1, Canada
[4]   School of Public Health, University of Montreal, Montreal, QC H3N 1X9, Canada
[5]   Department of Community Medicine, Tehran University of Medical Sciences, Tehran 14117-13135, Iran
*   Correspondence: robert.turcotte.med@ssss.gouv.qc.ca

Simple Summary: A solitary fibrous tumor (SFT) is a fibroblastic mesenchymal tumor with the hallmark of an NAB2–STAT6 gene fusion and an intermediate tendency to metastasize. Based on the lack of a histologic-based grading system for extra-meningeal SFTs, we defined the prognostic value of histologic features that predict the risk of developing distant metastases. Moreover, our study revealed the histologic alterations to recurrent SFTs that affect the biological behavior of the tumor.

Abstract: Histology has not been accepted as a valid predictor of the biological behavior of extra-meningeal solitary fibrous tumors (SFTs). Based on the lack of a histologic grading system, a risk stratification model is accepted by the WHO to predict the risk of metastasis; however, the model shows some limitations to predict the aggressive behavior of a low-risk/benign-appearing tumor. We conducted a retrospective study based on medical records of 51 primary extra-meningeal SFT patients treated surgically with a median follow-up of 60 months. Tumor size ($p = 0.001$), mitotic activity ($p = 0.003$), and cellular variants ($p = 0.001$) were statistically associated with the development of distant metastases. In cox regression analysis for metastasis outcome, a one-centimeter increment in tumor size enhanced the expected metastasis hazard by 21% during the follow-up time (HR = 1.21, CI 95% (1.08–1.35)), and each increase in the number of mitotic figures escalated the expected hazard of metastasis by 20% (HR = 1.2, CI 95% (1.06–1.34)). Recurrent SFTs presented with higher mitotic activity and increased the likelihood of distant metastasis ($p = 0.003$, HR = 12.68, CI 95% (2.31–69.5)). All SFTs with focal dedifferentiation developed metastases during follow-up. Our findings also revealed that assembling risk models based on a diagnostic biopsy underestimated the probability of developing metastasis in extra-meningeal SFTs.

Keywords: solitary fibrous tumor; extrameningeal; dedifferentiation; cellular variant

Citation: Hassani, M.; Jung, S.; Ghodsi, E.; Seddigh, L.; Kooner, P.; Aoude, A.; Turcotte, R. Value of Cellular Components and Focal Dedifferentiation to Predict the Risk of Metastasis in a Benign-Appearing Extra-Meningeal Solitary Fibrous Tumor: An Original Series from a Tertiary Sarcoma Center. *Cancers* **2023**, *15*, 1441. https://doi.org/10.3390/cancers15051441

Academic Editor: Bahil Ghanim

Received: 21 December 2022
Revised: 21 February 2023
Accepted: 22 February 2023
Published: 24 February 2023

1. Introduction

Solitary fibrous tumors (SFTs) are a subset of fibroblastic mesenchymal soft tissue tumors with the hallmark of an NAB2–STAT6 gene fusion [1–3] and present a metastatic rate of up to 34% after surgical resection [4,5]. The evolution of SFT histopathology has abolished some previous tumor entities, such as hemangiopericytoma; however, histology has not been accepted as a valid predictor of SFT biological behavior [2,6,7]. Many studies have tried to determine some criteria to differentiate benign and malignant variants of the tumor. Some suggested histologic features, including mitosis, atypia, and tumor necrosis, as well as demographic features, including age, tumor size, and location of the tumor as predictors of SFT aggressive behavior. Particularly size and mitotic counts have been emphasized more than others [5,8–12]. Based on the lack of a histologic grading system, Demicco et al. developed a risk stratification model based on the patient's age, tumor size, mitotic count, and tumor necrosis. The model is accepted by the World Health Organization (WHO) to predict the risk of metastasis in extra-meningeal SFT patients [7,13–15]. Not only a histologically benign-appearing SFT, but a low-risk SFT according to Demicco's model may present with distant metastases [11,16]. Furthermore, the patient's age, which is comparable to histologic items in Demicco's model, has inconsistent predictive value for the biological behavior of the tumor [4,11,16,17]. A histology-based grading system for SFTs originating from the central nervous system (CNS) is available whilst debating on the risk assessment of extra-meningeal SFTs [18,19]. The current study aimed to contribute to the debates on the biological behavior of extra-meningeal SFTs and to find some characteristics that predict the risk of metastases in a low-risk SFT.

2. Methods

2.1. Patients

Upon receiving approval from the Research Ethics Board (REB), we collected the medical records of 99 patients who were diagnosed with one of the SFT-related pathology reports, which included solitary fibrous tumor, hemangiopericytoma, anaplastic hemangiopericytoma, and malignant solitary fibrous tumor. The inclusion criterion was any patients with a primary extra-meningeal solitary fibrous tumor that was treated with surgical resection and followed within our tertiary sarcoma center. The diagnosis of SFT in our cases was based on histological features supported by ancillary immunohistochemical tests, such as STAT6, CD34, CD99, Vimentin, and BCL2. Any questionable diagnoses were reassessed by a sarcoma pathologist (S.J). The exclusion criteria were 1, SFT originating from the meningeal membrane or central nervous system (CNS), 2, metastatic lesions without a primary tumor histologic report, and 3, patients lacking appropriate follow-up data. We excluded 16 patients who had a tumor originating from the central nervous system (CNS); furthermore, 32 patients were excluded from the final statistical analysis because of having incomplete medical records or being diagnosed with metastasis at initial presentation without primary tumor pathology available. Ultimately, 51 primary extra-meningeal SFTs that were treated with surgical resection between 2005 to 2021 were included in the final analysis. The histologic characteristics of each tumor, such as frequency of mitosis per 10 high power fields (HPFs), a total necrotic area equal to or more than 10% of the tumor, having a hypercellular component, having a focal dedifferentiated nodule, and cyto-atypia, were extracted from the pathology reports of the first diagnostic biopsy and the resected tumor.

2.2. Statistical Analysis

Categorical variables included tumor necrosis, atypia, cellularity, the patient's gender, and the tumor's primary location. Numerical variables included the patient's age, tumor size, and mitosis count. Simultaneously, the age was stratified into <55 and ≥55 to be assessed as a categorical variable. Categorical variables were presented by frequency and valid percent (VP), and continuous variables were presented by median, range, and standard deviation (SD). The metastatic/recurrence risk score was calculated according to Demicco's [14] and Pasquali's [9] models for each patient. The en bloc resection of the primary tumor was considered index surgery. The outcomes were distant metastasis, which we considered as an indicator of aggressive behavior of the tumor, and the patient's five-year disease-free (5YDF) survival after the index surgery. As we excluded the patients with incomplete follow-up from our final analysis, we used the logistic regression model, Fisher exact test, and Chi-square test to address any statistical association between variables and distant metastases, using SPSS version 25 and Stata version 14. We also performed cox regression analysis and applied Kaplan–Meier survival curve analysis to estimate the probability of outcomes during the follow-up. Receiver operating characteristic (ROC) curve analysis was used to determine the optimal values with the highest sensitivity and specificity for each continuous variable correlated with outcomes.

3. Results

3.1. Patients

Nine patients (18%) had distant metastasis in the follow-up period after the index surgery. Seven patients (14%) presented with local recurrence, and 78% of our patients have not shown local or distant tumor recurrence during the first 5-year surveillance after the index surgery. No simultaneous metastatic lesion had been found at the time of the index surgery; however, only one case was suspicious as having lung metastasis at the time of the tumor diagnosis. Table 1 presents the patient demographic and histology data.

Table 1. Patient demographic and histology data. VP: valid percent, SD: standard deviation, HPF: high power field, N: number, m: months.

Variables	Median (Range)	N (VP%)	Mean (SD)
Age (Y)	59 (27–81)		
Gender			
male		27 (53%)	
female		24 (47%)	
Tumor location			
Extremity/intramuscular		20 (39%)	
Retroperitoneum/viscera		18 (35%)	
Intra-thoracic		13 (25%)	
Follow-up (m)	60 (11–120)		
Tumor size (cm)	5.5 (1.1–23)		7.61 (5.03)
Necrosis		11 (23.4%)	
Cellularity		17 (42.5%)	
Atypia		18 (39.1%)	
Mitosis (10HPF)	2.5 (0–22)		4.15 (4.9)
Demicco's risk score			
low		32 (62.7%)	
Intermediate		13 (25.5%)	
High		12 (32.4%)	
Pasquali's risk score (extra pleural SFT)			
Very low/low		27 (72.9)	
Intermediate		5 (13.5)	
High		12 (32.4)	

3.2. Distant Metastasis

Distant metastases were detected in nine (18%) of our patients. The median time to distant metastasis was 24 months (range: 6–72) (Figure 1).

Gender and age had no significant statistical association with SFT distant metastasis (p values were 0.47 and 0.20, respectively). No significance was reached by dichotomizing age into younger than 55 and older than 55 ($p = 0.82$) when looking at metastasis. SFT location had no significant association with distant metastasis ($p = 0.42$). Nonetheless, only one intra-thoracic tumor, which had a focal dedifferentiated component, presented with metastasis. Tumor size increased the risk of distant metastasis ($p = 0.001$). In cox regression analysis for metastasis outcome, a one-centimeter increment in tumor size enhanced the expected metastasis hazard by 21% during the follow-up time (HR = 1.21, CI 95% (1.08–1.35)). ROC curve analysis showed that the optimal cut-point for tumor size to distinguish metastatic from non-metastatic outcomes is 7.4 cm (area under the curve (AUC) = 0.85).

Our analysis demonstrated that the cellular variant of the tumor (Figure 2) was associated with distant metastasis ($p = 0.001$). The Kaplan–Meier cure showed considerable differences between the two groups regarding the presence of a cellular component (Figure 3). Atypia ($p = 0.12$) and necrosis ($p = 0.073$) had no significant association with metastasis in our patients. The mitotic count was correlated with distant metastasis ($p = 0.003$). During follow-up, each increase in the number of mitotic figures escalated the expected hazard of metastasis by 20% (HR = 1.2, CI 95% (1.06–1.34)). Our analysis showed that a threshold of 1.5 mitoses was associated with an increased metastatic rate, using ROC curve analysis (AUC = 0.82).

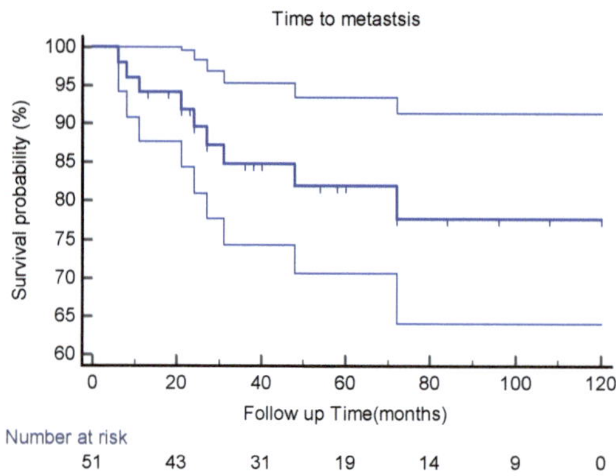

Figure 1. Kaplan–Meier curve estimates with 95%CIs for distant metastasis in extra-meningeal SFT patients.

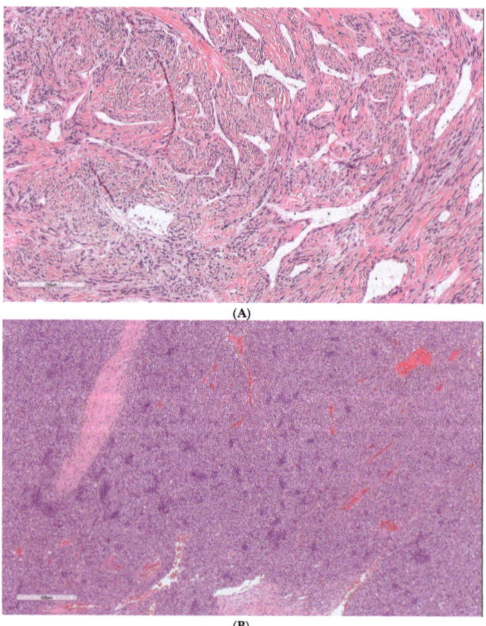

Figure 2. Solitary fibrous tumor H&E staining. The spectrum of cells and fibrous stroma in: (**A**) classic hyalinized solitary fibrous tumor and (**B**) cellular variant of the tumor. Cellular SFT is characterized by the tightly packed proliferation of ovoid to spindle cells arranged around conspicuous vessels and scant stromal components. The uncropped high-quality slides are shown in Supplementary File S1.

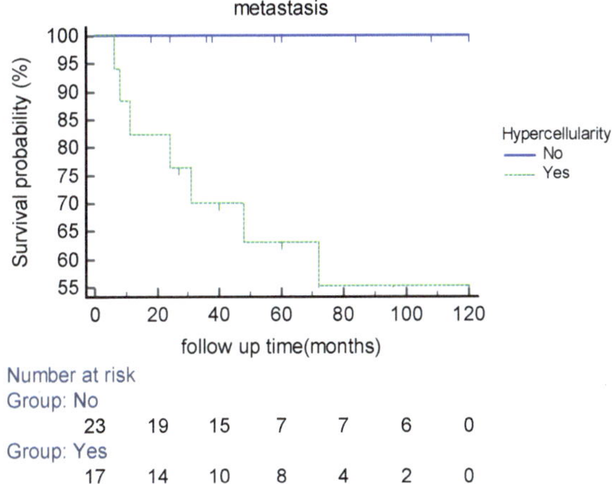

Figure 3. Kaplan–Meier curve estimates of the probability of distant metastasis in extra-meningeal SFT patients based on hypercellularity during the follow-up period.

Seven patients developed local recurrence. The mean mitotic counts per 10 HPFs in the primary and the recurrent lesions were 4.7 and 13.8, respectively. Tumor local recurrence was associated with a higher probability of developing distant metastasis in our patients ($p = 0.003$) (HR = 12.68, CI 95% (2.31–69.5)) (Figure 4).

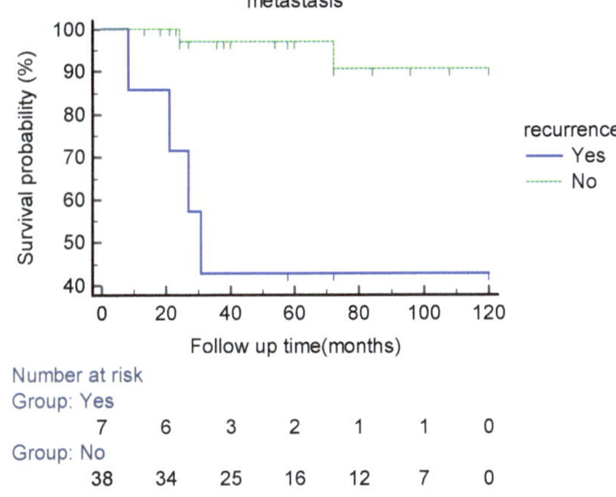

Figure 4. Kaplan–Meier curve estimates of the probability of distant metastasis in extra-meningeal SFT patients based on the local recurrence status during the follow up period.

3.3. Patient Survival

Thirty-one (78%) patients had no local recurrence or distant metastases during the 5-year follow-up after the index surgery, of which twelve had a post-index surgery surveillance of less than 60 months (Figure 5).

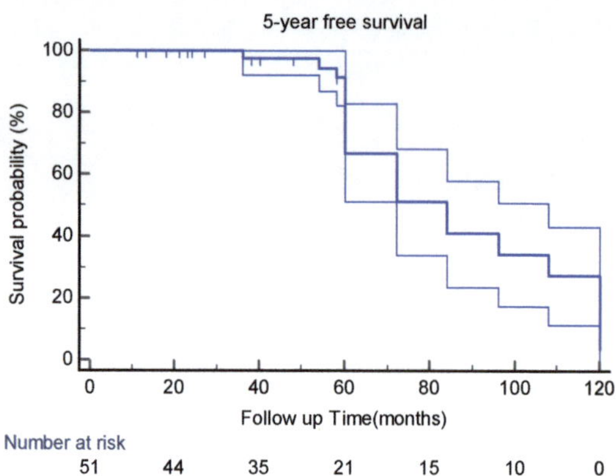

Figure 5. Kaplan–Meier curve estimates with 95%CIs for 5-year disease-free survival in extra-meningeal SFT patients.

In this study, the patient's age ($p = 0.12$) and gender ($p = 0.58$), tumor location ($p = 0.56$, 0.93 for different layers), cellular variant ($p = 0.41$), atypia ($p = 0.54$), and necrosis ($p = 0.98$) had no impact on the patient's 5YDF survival. Even the mitotic count had no statistical association with the patient's 5YDF survival ($p = 0.38$).

In the multivariable cox regression model, only the tumor size decreased the likelihood of having 5YDF survival in our patients ($p = 0.01$) (HR = 0.79, CI 95% (0.67–0.94)). Table 2 presents the statistical association between variables and outcomes.

Table 2. The statistical association between variables and outcomes are presented; HR: hazard ratio, CI: confidence interval, 5YDF: 5-year disease-free.

Variables	Metastasis	HR (CI 95%)	5YDF Survival	HR (CI 95%)
Age (continuous)	$p = 0.20$		$p = 0.12$	
Age (categorical)	$p = 0.82$		$p = 0.45$	
Hypercellularity	$p = 0.001$		$p = 0.41$	
Size	$p = 0.001$	1.21 (1.08–1.35)	$p = 0.01$	0.79 (0.67–0.94)
Mitotic activity	$p = 0.003$	1.2 (1.06–1.34)	$p = 0.38$	
Demicco score	$p = 0.007$	1.62 (1.14–2.3)	$p = 0.019$	1.63 (1.08–2.45)
Local recurrence	$p = 0.003$	12.68 (2.31–69.5)		

3.4. Risk Assessment Tools

The cumulative risk score based on Demicco's model was linked to a higher chance of developing distant metastasis ($p = 0.007$, HR = 1.62) and decreased the likelihood of having 5YDF survival ($p = 0.019$, HR = 1.63), using the pathologic report after the primary tumor resection. Categorical Demicco's score based on risk stratification groups also had a statistical relationship with metastasis and patient 5YDF survival (p values were 0.01, 0.004 respectively). Nevertheless, two patients from the low-risk group presented with distant metastases, and three patients from the high-risk group had no metastases during their follow-up (range 13–120 months).

In extra-plural SFTs, categorical Pasquali's risk score based on the histology of the index surgery was associated with distant metastasis ($p = 0.051$); however, it did not show statistical significance based on 5YDF survival ($p = 0.063$).

The risk score calculation based on diagnostic biopsy histology failed to predict the likelihood of distant metastasis in our patients, using both Demicco's ($p = 0.81$) and Pasquali's ($p = 0.15$) models.

3.5. Dedifferentiated SFT

Dedifferentiation in our cases was defined by a high-grade sarcoma in an abrupt transition from typical SFT (Figure 6). A focal dedifferentiated component was detected in four primary SFTs (7.8%), which were resected with negative surgical margins. All these patients presented with distant metastasis during the surveillance after the resection (range 21–48 months). Two tumors had a cellular component present in their diagnostic biopsy samples (Table 3).

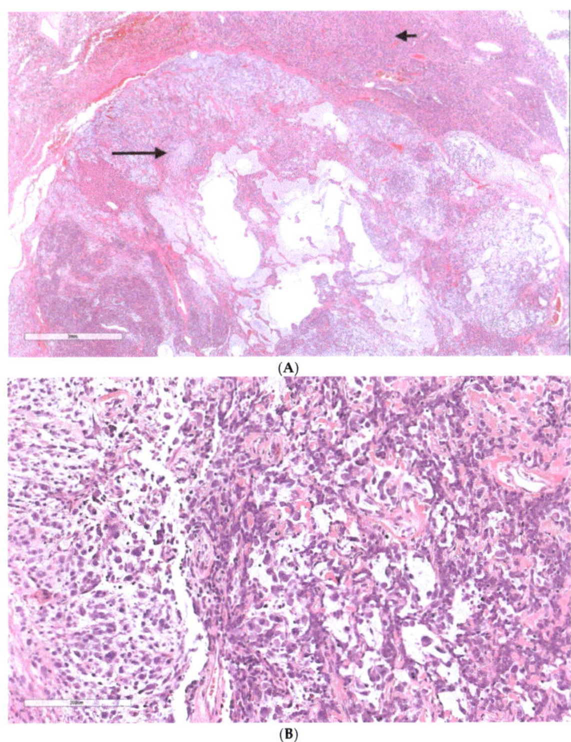

Figure 6. Dedifferentiated solitary fibrous tumor H&E staining. (**A**) Two distinct areas of the typical solitary fibrous tumor (upper part/short arrow) and high-grade pleomorphic sarcoma (lower part/long arrow). (**B**) High-grade undifferentiated sarcoma. The uncropped high-quality slides are shown in Supplementary File S1.

Table 3. Characteristics of extra-meningeal SFT patients with a dedifferentiated component. ND: not determined. Int: intermediate. HPF: high power field.

Case	Age	Location	Size cm	Mitosis 10HPF	Hyper Cellular	Demicco Risk Group	Dedifferentiated Nodule Histology
1	31	Paraspinal muscle	7.5	2	Yes	Low	Scattered epithelioid/rhabdoid change, negative CD34
2	59	Pre-sacral	14	10	ND	Int.	Focal high-grade dedifferentiated sarcoma, negative CD34
3	56	Pleura	11	13	Yes	Int.	Focal rhabdoid change
4	41	Paraspinal muscle	8.5	5	ND	Low	Focal myxoid/epithelioid change, negative CD34

4. Discussion

Histology grading has been accepted as a principal predictor of metastatic behavior in many soft tissue sarcomas [20,21]; however, there is no consensus on a histologic-based grading system for extra-meningeal SFT, which encompasses a spectrum of tumors from dedifferentiated variants to benign-appearing SFTs. Moreover, Demicco's risk stratification model has been accepted as a reliable statistical prognostic measure by the WHO, but it shows some limitations in predicting the aggressive behavior of the tumor [16]. Similarly, a few patients in our study, being classified as low risk due to Demicco's model, experienced distant metastases.

Age is a prognostic factor in the Demicco model; however, our study found that age as either a numerical or categorical variable had no significant association with the risk of developing metastases. On the contrary, hypercellularity as a histologic feature was reported to increase the odds of developing metastasis in previous studies, of which, some had the drawback of including meningeal SFTs [9,10,22,23]. Comparably, our study revealed that the cellular variant of the tumor in extra-meningeal SFT increased the risk of distant metastasis. Having no definite cut-point, cellularity is considered a subjective finding by critics [5]. Furthermore, it has not been considered a prognostic factor in the FNCLCC grading system. Our findings, however, strongly suggest that cellularity increases the prognostic value of risk assessment models much more than age, which will be aligned with the meningeal/CNS SFT grading system [18,19].

The current study revealed that we will underestimate the metastatic potential of an SFT if we assemble a risk model based on core needle biopsy information. To overcome this limitation, we must consider any prognostic factor that independently impacts the biological behavior of an SFT, to devise our treatment plan.

Tumor size was reported as a prognostic factor by many studies, and different cut-point values ranging from 8 to 15 cm have been calculated [8,17,24]. According to our findings, a tumor size larger than 7 cm, which is associated with higher odds of metastasis and independently impacts the patient's survival, should be approached as a tumor with a high probability of aggressive behavior.

Mitotic count was also associated with metastasis and impacted patient survival, either as an independent factor or as a part of risk assessment models in many studies [4,5,8,9]. Our findings, similarly, supported the prognostic value of mitotic activity ($\geq$2 mitosis/10 HPFs) in extra meningeal SFTs.

A solitary fibrous tumor can present with a synchronous focal dedifferentiated area [25,26]. In our patients, any primary tumor with a focal dedifferentiated nodule was associated with distant metastasis, albeit being surrounded by a benign-appearing SFT, and developing metastasis was independent of their negative surgical margins. Half of our patients with focal dedifferentiation were reported as having a cellular variant of the tumor in their diagnostic biopsy. Similar findings were revealed by other studies, which highlighted the aggressive behavior of focal dedifferentiated tumors, as well as the association between hypercellularity and the focal dedifferentiation of SFTs [23,27].

Our recurrent tumors showed more aggressive histologic features, such as higher mitotic rates. We also found a significant association between the risk of developing distant metastasis and local recurrence, albeit not being an independent variable. These important findings favored the aggressive treatment approach to an SFT with any histologic or demographic worrisome features; however, we need further studies to determine independent predictors for the above-mentioned finding.

Regarding the flourishing role of molecular studies in the diagnosis of soft tissue sarcomas [7,28,29], one study replaced the age with the MIB-1 proliferation index in their risk assessment model to predict the outcomes of SFT patients [30]. We assume that the molecular study overcomes the current flaws of predicting the biological behavior of an extra-meningeal SFT.

Our retrospective study had some inherent limitations: we deliberately excluded any patient with incomplete medical records, which might increase the risk of selection bias in our study. Considering the low frequency of outcomes in our five-year follow-up period, the results of cox regression analysis should be interpreted with caution; furthermore, it restricted the use of multivariate analysis to determine independent associations between variables and outcomes. In addition, half of our patients were followed for less than 5 years and may present with a metastatic lesion in the future considering the risk of late metastasis with SFTs [16,17]. Finally, we excluded molecular findings from our variables due to the lack of a complete molecular profile for all patients.

5. Conclusions

Hypercellular components and focal dedifferentiation predict the risk of developing metastases in extra-meningeal SFTs. Assembling a risk model based on biopsy features underestimates the metastatic potential of the tumor. Recurrent SFTs present with aggressive histologic features, such as higher mitotic activity, and increase the likelihood of distant metastasis.

Supplementary Materials: The following supporting information can be downloaded at: https://www.mdpi.com/article/10.3390/cancers15051441/s1, File S1: The uncropped high-quality slides are shown in Supplementary File S1.

Author Contributions: Conceptualization, R.T., S.J. and M.H.; methodology, E.G.; formal analysis, L.S.; software, L.S.; validation, E.G.; resources, S.J.; data curation, P.K.; writing—original draft preparation, M.H.; writing—review and editing, A.A. and P.K.; supervision, R.T. and A.A. All authors have read and agreed to the published version of the manuscript.

Funding: This study received no external funding.

Institutional Review Board Statement: The study was conducted according to the guidelines of the Declaration of Helsinki, and it has received approval from the Research Ethics Board of McGill University Health Centre (MUHC Authorization (SFT-clinical/2019-5006 and 2023-9404)).

Informed Consent Statement: This retrospective study did not present any specific name, personal details, patient photos, or other identifiable human material or data. The patients previously consented that their data can be used for future research studies.

Data Availability Statement: Upon reasonable request, the study data and analyses are available from the first author.

Acknowledgments: We would like to thank Nadine Zablith for her assistance in collecting data. The first author (M.H.) would like to acknowledge a fellowship award from the Cedar Cancer Foundation.

Conflicts of Interest: The authors declare no conflict of interest.

References

1. Doyle, L.A.; Vivero, M.; Fletcher, C.D.; Mertens, F.; Hornick, J.L. Nuclear expression of STAT6 distinguishes solitary fibrous tumor from histologic mimics. *Mod. Pathol.* **2014**, *27*, 390–395. [CrossRef] [PubMed]
2. Jo, V.Y.; Fletcher, C.D. WHO classification of soft tissue tumours: An update based on the 2013 (4th) edition. *Pathology* **2014**, *46*, 95–104. [CrossRef] [PubMed]
3. Huang, S.C.; Huang, H.Y. Solitary fibrous tumor: An evolving and unifying entity with unsettled issues. *Histol. Histopathol.* **2019**, *34*, 313–334. [CrossRef] [PubMed]
4. van Houdt, W.J.; Westerveld, C.M.; Vrijenhoek, J.E.; van Gorp, J.; van Coevorden, F.; Verhoef, C.; van Dalen, T. Prognosis of solitary fibrous tumors: A multicenter study. *Ann. Surg. Oncol.* **2013**, *20*, 4090–4095. [CrossRef] [PubMed]
5. Salas, S.; Resseguier, N.; Blay, J.-Y.; Le Cesne, A.; Italiano, A.; Chevreau, C.; Rosset, P.; Isambert, N.; Soulie, P.; Cupissol, D. Prediction of local and metastatic recurrence in solitary fibrous tumor: Construction of a risk calculator in a multicenter cohort from the French Sarcoma Group (FSG) database. *Ann. Oncol.* **2017**, *28*, 1779–1787. [CrossRef]
6. Fletcher, C.D. The evolving classification of soft tissue tumours: An update based on the new WHO classification. *Histopathology* **2006**, *48*, 3–12. [CrossRef]
7. Cloutier, J.M.; Charville, G.W. Diagnostic classification of soft tissue malignancies: A review and update from a surgical pathology perspective. *Curr. Probl. Cancer* **2019**, *43*, 250–272. [CrossRef]
8. Demicco, E.G.; Park, M.S.; Araujo, D.M.; Fox, P.S.; Bassett, R.L.; Pollock, R.E.; Lazar, A.J.; Wang, W.L. Solitary fibrous tumor: A clinicopathological study of 110 cases and proposed risk assessment model. *Mod. Pathol.* **2012**, *25*, 1298–1306. [CrossRef]

9. Pasquali, S.; Gronchi, A.; Strauss, D.; Bonvalot, S.; Jeys, L.; Stacchiotti, S.; Hayes, A.; Honore, C.; Collini, P.; Renne, S.L. Resectable extra-pleural and extra-meningeal solitary fibrous tumours: A multi-centre prognostic study. *Eur. J. Surg. Oncol. (EJSO)* **2016**, *42*, 1064–1070. [CrossRef]

10. Hassani, M.; Jung, S.; Garzia, L.; Ghodsi, E.; Alcindor, T.; Turcotte, R.E. Aggressive Behavior Predictors in Solitary Fibrous Tumor Demographic, Clinical, and Histopathologic Characteristics of 81 Cases. *Ann. Surg. Oncol.* **2021**, *28*, 6861–6867. [CrossRef]

11. DeVito, N.; Henderson, E.; Han, G.; Reed, D.; Bui, M.M.; Lavey, R.; Robinson, L.; Zager, J.S.; Gonzalez, R.J.; Sondak, V.K.; et al. Clinical Characteristics and Outcomes for Solitary Fibrous Tumor (SFT): A Single Center Experience. *PLoS ONE* **2015**, *10*, e0140362. [CrossRef] [PubMed]

12. Gengler, C.; Guillou, L. Solitary fibrous tumour and haemangiopericytoma: Evolution of a concept. *Histopathology* **2006**, *48*, 63–74. [CrossRef]

13. Sbaraglia, M.; Bellan, E.; Dei Tos, A.P. The 2020 WHO Classification of Soft Tissue Tumours: News and perspectives. *Pathologica* **2021**, *113*, 70–84. [CrossRef] [PubMed]

14. Demicco, E.G.; Wagner, M.J.; Maki, R.G.; Gupta, V.; Iofin, I.; Lazar, A.J.; Wang, W.L. Risk assessment in solitary fibrous tumors: Validation and refinement of a risk stratification model. *Mod. Pathol.* **2017**, *30*, 1433–1442. [CrossRef]

15. WHO Classification of Tumours Editorial Board. *Soft Tissue and Bone Tumours WHO Classification of Tumours*, 5th ed.; International Agency for Research on Cancer: Lyon, France, 2020; Volume 3.

16. Machado, I.; Nieto Morales, M.G.; Cruz, J.; Lavernia, J.; Giner, F.; Navarro, S.; Ferrandez, A.; Llombart-Bosch, A. Solitary Fibrous Tumor: Integration of Clinical, Morphologic, Immunohistochemical and Molecular Findings in Risk Stratification and Classification May Better Predict Patient outcome. *Int. J. Mol. Sci.* **2021**, *22*, 9423. [CrossRef] [PubMed]

17. Gholami, S.; Cassidy, M.R.; Kirane, A.; Kuk, D.; Zanchelli, B.; Antonescu, C.R.; Singer, S.; Brennan, M. Size and Location are the Most Important Risk Factors for Malignant Behavior in Resected Solitary Fibrous Tumors. *Ann. Surg. Oncol.* **2017**, *24*, 3865–3871. [CrossRef]

18. Louis, D.N.; Perry, A.; Reifenberger, G.; von Deimling, A.; Figarella-Branger, D.; Cavenee, W.K.; Ohgaki, H.; Wiestler, O.D.; Kleihues, P.; Ellison, D.W. The 2016 World Health Organization Classification of Tumors of the Central Nervous System: A summary. *Acta Neuropathol.* **2016**, *131*, 803–820. [CrossRef]

19. Louis, D.N.; Perry, A.; Wesseling, P.; Brat, D.J.; Cree, I.A.; Figarella-Branger, D.; Hawkins, C.; Ng, H.K.; Pfister, S.M.; Reifenberger, G.; et al. The 2021 WHO Classification of Tumors of the Central Nervous System: A summary. *Neuro. Oncol.* **2021**, *23*, 1231–1251. [CrossRef]

20. Coindre, J.M.; Terrier, P.; Guillou, L.; Le Doussal, V.; Collin, F.; Ranchère, D.; Sastre, X.; Vilain, M.O.; Bonichon, F.; N'Guyen Bui, B. Predictive value of grade for metastasis development in the main histologic types of adult soft tissue sarcomas: A study of 1240 patients from the French Federation of Cancer Centers Sarcoma Group. *Cancer Interdiscip. Int. J. Am. Cancer Soc.* **2001**, *91*, 1914–1926. [CrossRef]

21. Coindre, J.M. Grading of soft tissue sarcomas: Review and update. *Arch. Pathol. Lab. Med.* **2006**, *130*, 1448–1453. [CrossRef]

22. Kim, J.M.; Choi, Y.-L.; Kim, Y.J.; Park, H.K. Comparison and evaluation of risk factors for meningeal, pleural, and extrapleural solitary fibrous tumors: A clinicopathological study of 92 cases confirmed by STAT6 immunohistochemical staining. *Pathol.-Res. Pract.* **2017**, *213*, 619–625. [CrossRef] [PubMed]

23. Yamada, Y.; Kohashi, K.; Kinoshita, I.; Yamamoto, H.; Iwasaki, T.; Yoshimoto, M.; Ishihara, S.; Toda, Y.; Ito, Y.; Kuma, Y.; et al. Histological background of dedifferentiated solitary fibrous tumour. *J. Clin. Pathol.* **2022**, *75*, 397–403. [CrossRef] [PubMed]

24. Gold, J.S.; Antonescu, C.R.; Hajdu, C.; Ferrone, C.R.; Hussain, M.; Lewis, J.J.; Brennan, M.F.; Coit, D.G. Clinicopathologic correlates of solitary fibrous tumors. *Cancer* **2002**, *94*, 1057–1068. [CrossRef] [PubMed]

25. Collini, P.; Negri, T.; Barisella, M.; Palassini, E.; Tarantino, E.; Pastorino, U.; Gronchi, A.; Stacchiotti, S.; Pilotti, S. High-grade sarcomatous overgrowth in solitary fibrous tumors: A clinicopathologic study of 10 cases. *Am. J. Surg. Pathol.* **2012**, *36*, 1202–1215. [CrossRef] [PubMed]

26. Kallen, M.E.; Hornick, J.L. The 2020 WHO Classification: What's New in Soft Tissue Tumor Pathology? *Am. J. Surg. Pathol.* **2021**, *45*, e1–e23. [CrossRef]

27. Mosquera, J.M.; Fletcher, C.D. Expanding the spectrum of malignant progression in solitary fibrous tumors: A study of 8 cases with a discrete anaplastic component–is this dedifferentiated SFT? *Am. J. Surg. Pathol.* **2009**, *33*, 1314–1321. [CrossRef]

28. Martin-Broto, J.; Mondaza-Hernandez, J.L.; Moura, D.S.; Hindi, N. A Comprehensive Review on Solitary Fibrous Tumor: New Insights for New Horizons. *Cancers* **2021**, *13*, 2913. [CrossRef] [PubMed]

29. Ouladan, S.; Trautmann, M.; Orouji, E.; Hartmann, W.; Huss, S.; Büttner, R.; Wardelmann, E. Differential diagnosis of solitary fibrous tumors: A study of 454 soft tissue tumors indicating the diagnostic value of nuclear STAT6 relocation and ALDH1 expression combined with in situ proximity ligation assay. *Int. J. Oncol.* **2015**, *46*, 2595–2605. [CrossRef]

30. Diebold, M.; Soltermann, A.; Hottinger, S.; Haile, S.R.; Bubendorf, L.; Komminoth, P.; Jochum, W.; Grobholz, R.; Theegarten, D.; Berezowska, S.; et al. Prognostic value of MIB-1 proliferation index in solitary fibrous tumors of the pleura implemented in a new score—A multicenter study. *Respir Res.* **2017**, *18*, 210. [CrossRef]

Article

Risk Stratification for Management of Solitary Fibrous Tumor/Hemangiopericytoma of the Central Nervous System

Connor J. Kinslow [1,2], Ali I. Rae [3], Prashanth Kumar [1], Guy M. McKhann [2,4], Michael B. Sisti [2,4], Jeffrey N. Bruce [2,4], James B. Yu [1,2], Simon K. Cheng [1,2,*] and Tony J. C. Wang [1,2,*]

1   Department of Radiation Oncology, Vagelos College of Physicians and Surgeons, Columbia University Irving Medical Center, 622 West 168th Street, BNH B011, New York, NY 10032, USA
2   Herbert Irving Comprehensive Cancer Center, Vagelos College of Physicians and Surgeons, Columbia University Irving Medical Center, 1130 St Nicholas Ave, New York, NY 10032, USA
3   Department of Neurological Surgery, Oregon Health & Sciences University, 3181 SW Sam Jackson Pkwy, Portland, OR 97239, USA
4   Department of Neurosurgery, Vagelos College of Physicians and Surgeons, Columbia University Irving Medical Center, 710 West 168th Street, New York, NY 10032, USA
*   Correspondence: sc3225@cumc.columbia.edu (S.K.C.); tjw2117@cumc.columbia.edu (T.J.C.W.); Tel.: +1-212-305-7077 (S.K.C. & T.J.C.W.); Fax: +1-212-305-5935 (S.K.C. & T.J.C.W.)

Simple Summary: A solitary fibrous tumor (SFT)/hemangiopericytoma (HPC) of the central nervous system (CNS) represents a rare meningeal tumor with the propensity to recur almost invariably and to metastasize extracranially. Given the rarity of the disease, there are no prospective trials by which to guide its management, and indications for radiotherapy are unclear. The NRG Oncology and European Organization for Research and Treatment of Cancer (EORTC) cooperative groups recently completed the first prospective trials to evaluate risk-adapted radiotherapeutic strategies for meningiomas, based on tumor grade and extent of resection. Using a similar approach, we created three risk categories using two large national US datasets. Our risk categories were highly prognostic of overall and cause-specific survival. Furthermore, our risk categories predicted the survival benefit associated with radiotherapy, which was limited to the high-risk group and, potentially, the intermediate-risk group. Our data suggest that a risk-adapted approach may be employed for the management of SFT/HPC of the CNS. These risk categories may be used in future retrospective and/or prospective studies.

Abstract: Introduction: Solitary fibrous tumor/hemangiopericytoma (SFT/HPC) of the central nervous system (CNS) is a rare meningeal tumor. Given the absence of prospective or randomized data, there are no standard indications for radiotherapy. Recently, the NRG Oncology and EORTC cooperative groups successfully accrued and completed the first prospective trials evaluating risk-adapted adjuvant radiotherapy strategies for meningiomas. Using a similar framework, we sought to develop prognostic risk categories that may predict the survival benefit associated with radiotherapy, using two large national datasets. Methods: We queried the National Cancer Database (NCDB) and the Surveillance, Epidemiology, and End Results (SEER) databases for all newly diagnosed cases of SFT/HPC within the CNS. Risk categories were created, as follows: low risk—grade 1, with any extent of resection (EOR) and grade 2, with gross–total resection; intermediate risk—grade 2, with biopsy/subtotal resection; high risk—grade 3 with any EOR. The Kaplan–Meier method and Cox proportional hazards regressions were used to determine the association of risk categories with overall and cause-specific survival. We then determined the association of radiotherapy with overall survival in the NCDB, stratified by risk group. Results: We identified 866 and 683 patients from the NCDB and SEER databases who were evaluated, respectively. In the NCDB, the 75% survival times for low- (n = 312), intermediate- (n = 239), and high-risk (n = 315) patients were not reached, 86 months (HR 1.60 (95% CI 1.01–2.55)), and 55 months (HR 2.56 (95% CI 1.68–3.89)), respectively. Our risk categories were validated for overall and cause-specific survival in the SEER dataset. Radiotherapy was associated with improved survival in the high- (HR 0.46 (0.29–0.74)) and intermediate-risk groups

Citation: Kinslow, C.J.; Rae, A.I.; Kumar, P.; McKhann, G.M.; Sisti, M.B.; Bruce, J.N.; Yu, J.B.; Cheng, S.K.; Wang, T.J.C. Risk Stratification for Management of Solitary Fibrous Tumor/Hemangiopericytoma of the Central Nervous System. *Cancers* **2023**, *15*, 876. https://doi.org/10.3390/cancers15030876

Academic Editor: Bahil Ghanim

Received: 23 December 2022
Revised: 19 January 2023
Accepted: 21 January 2023
Published: 31 January 2023

(HR 0.52 (0.27–0.99)) but not in the low-risk group (HR 1.26 (0.60–2.65)). The association of radiotherapy with overall survival remained significant in the multivariable analysis for the high-risk group (HR 0.55 (0.34–0.89)) but not for the intermediate-risk group (HR 0.74 (0.38–1.47)). Similar results were observed in a time-dependent landmark sensitivity analysis. Conclusion: Risk stratification based on grade and EOR is prognostic of overall and cause-specific survival for SFT/HPCs of the CNS and performs better than any individual clinical factor. These risk categories appear to predict the survival benefit from radiotherapy, which is limited to the high-risk group and, potentially, the intermediate-risk group. These data may serve as the basis for a prospective study evaluating the management of meningeal SFT/HPCs.

Keywords: solitary fibrous tumor; hemangiopericytoma; radiotherapy; risk stratification

1. Introduction

Solitary fibrous tumor (SFT)/hemangiopericytoma (HPC) of the central nervous system (CNS) is a rare meningeal tumor with an incidence rate of 3.8 cases per 10,000,000 persons per year in the US, which is rising [1,2]. The incidence rate approached 6 persons per 10,000,000 persons per year in 2013, or approximately 230 cases diagnosed annually. In 2016, the World Health Organization (WHO) created a combined designation of SFT/HPC, recognizing that the two tumors share the *NAB2/STAT6* fusion and, therefore, likely represent tumors with a common genetic etiology along a spectrum of possible clinical behaviors [3]. Unlike meningiomas and low-grade solitary fibrous tumors [4–6], hemangiopericytomas recur almost invariably [7–15] and have a high propensity for extracranial metastasis [15]. The most recent WHO update in November 2021 (CNS-5) removed the term "hemangiopericytoma" so that the tumor name would conform fully with soft-tissue pathology nomenclature [16].

Optimal management of CNS SFT/HPC includes maximal safe resection, with or without adjuvant radiotherapy [1]. Because there are no randomized controlled trials or prospective studies, indications for adjuvant radiotherapy remain unclear and are institution-dependent. Adjuvant radiotherapy is administered for approximately 53% of HPCs classified as grades 2–3 in the US [1,17]. Retrospective series, population-based studies, and meta-analyses have yielded mixed results regarding the survival benefit of radiotherapy, likely due to selection bias and confounding clinical factors [8,9,13–15,18–28]. Adjuvant radiotherapy is more likely to benefit patients with higher-grade tumors or a lesser extent of resection (EOR) [1]; however, there is no consensus on the absolute indications. Retrospective real-world datasets are unlikely to simulate clinical trial outcomes [29]. Prospective studies to help guide management are, therefore, vitally needed.

NRG Oncology (formerly known as the Radiation Therapy Oncology Group (RTOG)) and the European Organization for Research and Treatment of Cancer (EORTC) have now both successfully enrolled and completed the first prospective, non-randomized phase-II trials evaluating adjuvant radiotherapeutic strategies for meningiomas [30–34]. These trials have been used to create risk-adapted standardized treatments and are also the basis for ongoing randomized clinical trials, thus laying the groundwork for evidence-based management of meningiomas.

No similar risk-adapted strategies have been developed for SFT/HPC. Here, we propose a risk-stratification schema for SFT/HPC, which may be considered as a foundation for future prospective or retrospective studies, with the intention of developing more standardized treatment paradigms. Similar to the RTOG and EORTC trials, we formulated prognostic risk groups based on tumor grade and EOR. We hypothesized that risk stratification could model prognosis better than any one individual clinical feature and thereby predict the survival benefit from radiotherapy. As a result, we analyzed risk categories and treatment-related outcomes reported in two large national databases.

2. Materials and Methods

2.1. Data Sources

The National Cancer Database (NCDB) is a retrospective nationwide dataset sponsored by the American College of Surgeons and the American Cancer Society, constituting 70% of invasive cancer cases diagnosed in the United States. Data were collected at over 1500 Commission on Cancer–accredited hospitals between 2004 and 2018 [35]. This database has been validated for several variables [36–39].

The Surveillance, Epidemiology, and End Results (SEER) program is the National Cancer Institute's (NCI) authoritative source for data on cancer incidence and survival [40]. It is considered the gold standard for cancer data collection internationally [36]. The SEER 18 database is populated with data from national cancer registries in 13 states, covering approximately 27.8% of the United States population [40]. The Commission on Cancer of the American College of Surgeons requires the participating cancer registries to collect information on malignancies that are diagnosed and/or treated at the hospital. Vital status is updated annually and the database routinely undergoes quality-control checks. Our methodology was conducted as described previously [41–47].

2.2. Patient Selection and Coding

We queried the NCDB (2018 submission) to identify cases of SFT (International Classification of Diseases (ICD)-O-3 code 8815) and HPC (ICD-O-3 code 9150) within the CNS (ICD-O-3 codes C70.1–C72.9) diagnosed between 1 January 2004 and 31 December 2016. The last possible date of follow-up for all cases was 31 December 2018. The following variables were collected and coded: age at diagnosis, sex, race, Charlson–Deyo score, primary site, tumor size, ICD-O-3 histology, ICD-0-3 behavior, collaborative staging (CS) site-specific factor 1 (WHO grade), surgery at the primary site, and radiation therapy.

Grades were determined using all the available information from ICD-O-3 histology, ICD-0-3 behavior, and CS site-specific factor 1 (WHO grade), to keep them consistent with the WHO 2016 grading criteria. All primary tumors reported to US cancer registries contain both a 4-digit ICD-0-3 histology code and a 5th digit for ICD-0-3 behavior. Behavior coding is based on histological morphology and indicates the likely behavior of the tumor in terms of its potential to invade the surrounding tissue, based on the behavior that most pathologists believe is usual for that tumor type. ICD-0-3 behavior coding can be changed at the discretion of the coding pathologist. Tumors are classified as benign, borderline malignant, or malignant. Tumors are coded as borderline malignant based on a pathologist's observations that the tumor has "low, borderline, or uncertain malignant potential". Based on the WHO 2016 grading criteria, SFTs were coded as grade 1 and HPCs as grade 2, unless the tumors displayed malignant behavior, in which case they were coded as grade 3. This was compared with the WHO grade when it was available, and the findings were generally concordant. In cases where the histology/behavior codes were discordant with the WHO grade, the WHO grade was used. Information on molecular analysis, including STAT6 immunostaining and/or *NAB2-STAT6*, was not available.

The extent of resection was based on definitions in the American College of Surgeons Commission on Cancer's Facility Oncology Registry Data System (FORDS) manual [48]. Primary site surgeries in US cancer registries are defined as "cancer-directed" if the goal of treatment is to modify, control, remove, or destroy cancer tissue. Incisional biopsies are not considered to be cancer-directed surgeries. Most patients that did not undergo cancer-directed surgeries had received histological confirmation of disease and were, therefore, assumed to have undergone biopsy. EOR was coded as a biopsy/STR or GTR, based on surgery with the primary site variable: "no surgery" (code 00 (no surgery of the primary site)), "subtotal resection" (STR) (codes 10 (tumor destruction, not otherwise specified), 21 (STR), 20 (local excision or excisional biopsy), 22 (resection of the tumor in the spinal cord or nerve), 40 (partial resection of the lobe of the brain when surgery cannot be coded as 20–30)), and "gross-total resection" (GTR) codes (30 (radical, total, gross resection of the tumor) and 55 (GTR of a lobe of the brain)), as is consistent with prior studies [1,49–53]. Because surgical coding in cancer registries is based on the anatomical extent of the resection and not on the residual tumor, we performed multiple sensitivity analyses using different EOR coding schemas.

We combined the grade and EOR variables and then further grouped those cohorts of patients with similar overall survival prognoses. Risk categories were created as follows: low risk—grade 1 with any EOR, grade 2 with GTR; intermediate risk—grade 2 with biopsy/STR; high risk—grade 3 with any EOR.

Patients were excluded if follow-up time was less than two months as these patients either did live long enough to undergo adjuvant treatment or see an effect of management. We also excluded patients with metastatic disease or that could not be defined by our risk-stratification schema.

We also queried the SEER 17 database (November 2021 submission (2000–2019)) [54] for newly diagnosed cases of SFT/HPC that were diagnosed between 1 January 2000 and 31 December 2019, with follow-ups through December 2020. The following variables were collected and coded: age at diagnosis, sex, race, ICD-O-3 histology, ICD-0-3 behavior, primary site, surgery at the primary site, and collaborative staging (CS) site-specific factor 1 (WHO grade). Uniform coding across all years of analysis was not available for other potentially prognostic variables, such as size. Cases diagnosed at autopsy, or that could have 0 days of follow-up, and cases with less than 1 month of follow-up were excluded, as were cases that could not be defined by our risk-stratification classification.

2.3. Statistical Analysis

Median survival times were determined using the Kaplan–Meier method, and significance was determined using the log-rank test. The 75th percentile survival time was used as a surrogate marker of survival when the median survival time was not reached [49,50,55]. Univariable and multivariable analyses of both overall survival (OS) and cause-specific survival (CSS) were conducted using the Cox proportional hazards ratios model, with logistic regressions. The 95% confidence intervals were expressed next to the corresponding hazard ratios (HR). Tests with two-tailed p-values < 0.05 were considered to be statistically significant. Demographic and clinical features that were significantly associated with survival were included in the multivariable analyses. Statistical analyses were conducted using SEER*Stat, version 8.3.9 (National Cancer Institute, Bethesda, MD, USA) and RStudio version 1.4.1106 (R-Project for Statistical Computing, Boston, MA, USA) software.

3. Results

3.1. Patient Selection and Clinical/Demographic Characteristics

We identified 1,578 patients in the NCDB who were newly diagnosed with SFT or HPC. After excluding those patients with metastatic disease ($n = 32$), less than two months of follow-up ($n = 163$), or unknown EOR ($n = 479$), there were 866 patients available for analysis. The median follow-up time for all cases was 44 months, with 149 deaths. Demographic and clinical characteristics of the patient population are displayed in Table 1.

In SEER, there were 715 cases of SFT/HPC. After excluding those patients diagnosed at autopsy, who could be considered to have 0 days of follow-up or less than 1 month of follow-up ($n = 19$), or an unknown extent of resection ($n = 13$) were excluded, 683 cases were available for analysis. The median follow-up time was 66 months. There were 197 recorded deaths, 62 of which were attributed to SFT/HPC (Table S1 in the Supplementary Materials).

Table 1. Demographic and clinical characteristics of the NCDB cohort.

Characteristic	Low-Risk, $N = 312$ [1]	Intermediate-Risk, $N = 239$ [1]	High-Risk, $N = 315$ [1]
Age	54 (43, 65)	55 (43, 66)	54 (42, 66)
Sex			
Male	149 (48%)	112 (47%)	156 (50%)
Female	163 (52%)	127 (53%)	159 (50%)
Race			
White	252 (81%)	200 (84%)	267 (85%)
Black	25 (8.0%)	28 (12%)	26 (8.3%)
Other/Unknown	14 (4.5%)	4 (1.7%)	6 (1.9%)
Asian/Pacific Islander	20 (6.4%)	7 (2.9%)	16 (5.1%)
Unknown	1	0	0
Charlson–Deyo Comorbidity Index			
0	252 (81%)	187 (78%)	236 (75%)
1	46 (15%)	35 (15%)	47 (15%)
2 or more	14 (4.5%)	17 (7.1%)	32 (10%)
Site			
Brain	241 (77%)	168 (70%)	269 (85%)
Spinal/Other CNS	71 (23%)	71 (30%)	46 (15%)
Histology			
SFT	115 (37%)	0 (0%)	22 (7.0%)
HPC	197 (63%)	239 (100%)	293 (93%)
Grade			
G1	115 (37%)	0 (0%)	0 (0%)
G2	197 (63%)	239 (100%)	0 (0%)
G3	0 (0%)	0 (0%)	315 (100%)
Tumor Size			
5cm or less	128 (41%)	103 (43%)	118 (37%)
Greater than 5cm	110 (35%)	51 (21%)	91 (29%)
Unknown	74 (24%)	85 (36%)	106 (34%)
EOR			
No surgery/STR	72 (23%)	239 (100%)	163 (52%)
GTR	240 (77%)	0 (0%)	152 (48%)
Radiation			
No radiotherapy	209 (67%)	132 (56%)	93 (30%)
Radiotherapy	102 (33%)	104 (44%)	219 (70%)
Unknown	1	3	3
Follow-up Time	45 (29, 69)	49 (30, 74)	41 (26, 61)
Vital Status			
0	281 (90%)	195 (82%)	241 (77%)
1	31 (9.9%)	44 (18%)	74 (23%)

[1] Median (IQR); n (%).

3.2. Development of Risk Stratification Model

In NCDB, the median survival time for all patients was not reached, with a 75% survival time of 86 months. The 75% survival times for tumors of grades 1, 2, and 3 were 92, 89, and 55 months, respectively ($p = 0.001$) (see Figure S1 in the Supplementary Materials). Compared with grade 1, grade 3 (HR 2.53 (95% CI 1.37–4.66), $p = 0.003$) but not grade 2 (HR 1.34 (95% CI 0.72–2.48), $p = 0.36$) disease was associated with poorer survival rates. GTR was associated with an improved OS rate compared with biopsy/STR patients (75% survival time of 99 vs. 68 months, HR 0.59 (95% CI 0.42–0.84), $p = 0.003$).

We combined the grades and EOR to create the following risk categories: low risk—grade 1 with any EOR and grade 2 with GTR; intermediate risk—grade 2 with biopsy/STR; high risk—grade 3 with any EOR. These risk categories improved the prognostic value compared with any single risk factor. The 75% survival times for low-, intermediate-, and high-risk tumors were calculated as not reached, 86 months, and 55 months, respectively ($p < 0.001$,

Figure 1A). Compared with low-risk disease, intermediate-risk, and high-risk disease were associated with poorer OS on univariable (HR 1.60 (95% CI 1.01–2.55), *p* = 0.05 and HR 2.56 (95% CI 1.68–3.89), *p* < 0.001, respectively (see Table 2 and Table S2 in the Supplementary Materials)), and multivariable analysis (HR 1.52 (95% CI 0.95–2.41), *p* = 0.08 and HR 2.38 (95% CI 1.56–3.63), *p* < 0.001, respectively).

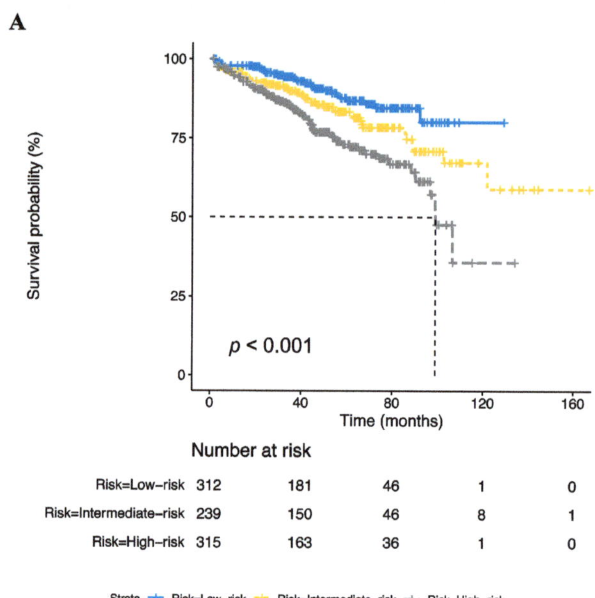

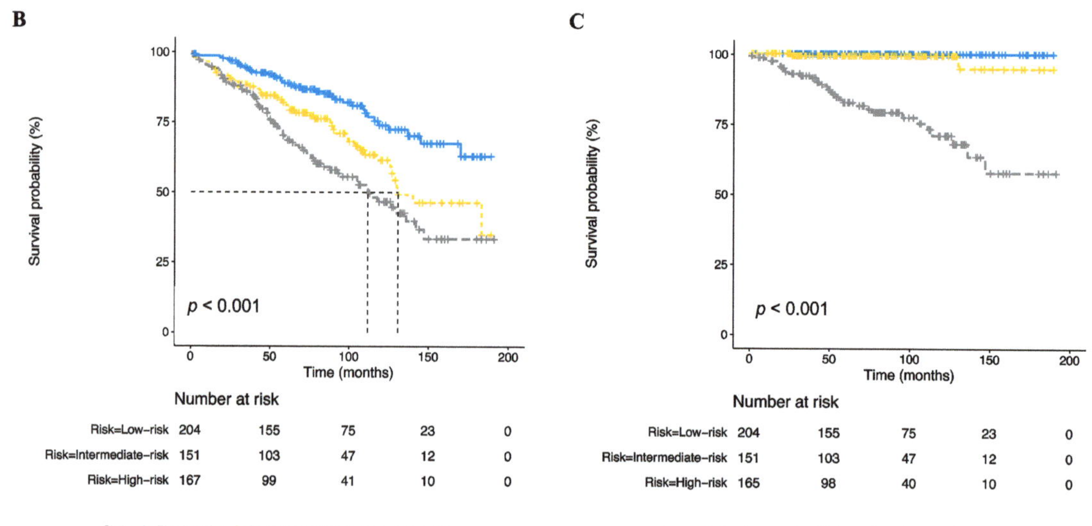

Figure 1. Kaplan–Meier curves representing overall survival in the NCDB (**A**) and overall (**B**) and cause-specific survival (**C**) in the SEER database based on risk groups.

Table 2. Summary of the univariable and multivariable analyses of the risk groups in NCDB and SEER.

Dataset/Characteristic	Univariable			Multivariable		
	HR	**95% CI**	*p*-Value	**HR**	**95% CI**	*p*-Value
NCDB [1]						
Low risk	-	-		-	-	
Intermediate risk	1.60	1.01, 2.55	**0.045**	1.52	0.95, 2.41	0.079
High risk	2.56	1.68, 3.89	**<0.001**	2.38	1.56, 3.63	**<0.001**
SEER [2]						
Low risk	-	-		-	-	
Intermediate risk	1.90	1.25, 2.90	**0.003**	1.94	1.27, 2.95	**0.002**
High risk	2.76	1.86, 4.08	**<0.001**	2.62	1.76, 3.92	**<0.001**

The table provides summary statistics from the univariable and multivariable analyses. Full univariable and multivariable analyses are included in the Supplementary Materials (Tables S2 and S3). Variables that were significant in the univariable analysis were included in the multivariable analyses. Bold values are statistically significant. [1] Variables included in the analysis were age, sex, race, the Charlson–Deyo comorbidity index, tumor size, anatomical site, risk group, and radiotherapy. [2] Variables included in the analysis were age, sex, race, anatomical site, and risk group.

We sought to validate our risk-stratification model in SEER. Although there may be an overlap of patients in SEER and NCDB, these databases use fundamentally distinct mechanisms to collect patient data; they undergo different quality-control processes and contain different variables. In the SEER dataset, risk stratification also improved the prognostic modeling over any single risk factor (Figure 1B and Figure S1 in the Supplementary Materials, $p < 0.001$). The 75% survival times for low-, intermediate-, and high-risk patients were 119, 88 (HR 1.90 (95% CI 1.25–2.90), $p = 0.003$), and 51 months (HR 2.76 (95% CI 1.86–4.08), $p < 0.001$), respectively (Table 2 and Table S3 in the Supplementary Materials). When evaluating CSS, risk stratification was also associated with improved prognostication, compared with individual clinical factors (Figure 1C). There were 0, 2 (1.3%), and 33 (20%) cause-specific deaths in the low-, intermediate-, and high-risk groups, respectively, with the corresponding 75% survival times not reached, not reached, and 111 months ($p < 0.001$), respectively. Given that there were no events in the low-risk group, the corresponding HRs could not be calculated.

3.3. Risk Stratification Predicts Benefit of Radiotherapy

Of the 859 patients with known radiotherapy status (99.2%), 425 (49%) received radiotherapy. Across all patients in the NCDB dataset, radiotherapy was not associated with improved OS (75% survival times of 89 vs. 73 months, HR 0.84 (0.61–1.17), $p = 0.30$, see Table S2 in the Supplementary Materials). However, when stratifying according to risk group, radiotherapy was associated with an improved OS in the high-risk (75% survival time 78 vs. 33 months, HR 0.46 (0.29–0.74), $p = 0.001$) and intermediate-risk groups (89 vs. 66 months, HR 0.52 (0.27–0.99), $p = 0.05$), but not in the low-risk group (not reached vs. not reached, HR 1.26 (0.60–2.65), $p = 0.55$, Figure 2, Table 3, and Tables S4–S6 in the Supplementary Materials). With the multivariable analysis, radiotherapy remained associated with an improved OS in the high-risk group (HR 0.59 (0.36–0.95), $p = 0.03$) but not in the intermediate-risk group (HR 0.74 (0.38–1.47), $p = 0.39$).

3.4. Sensitivity Analyses

ICD-O-3 histology and behavior were available for 100% of cases. Additional information on WHO grades was available for 558 (64.4%) patients in the NCDB. The grade of the tumor was modified for 126 (22.5%) cases when the WHO grade was included. We performed a sensitivity analysis, excluding those patients with missing data. Similar results were observed when we only included those patients with all histological data points (Figure S2 in the Supplementary Materials).

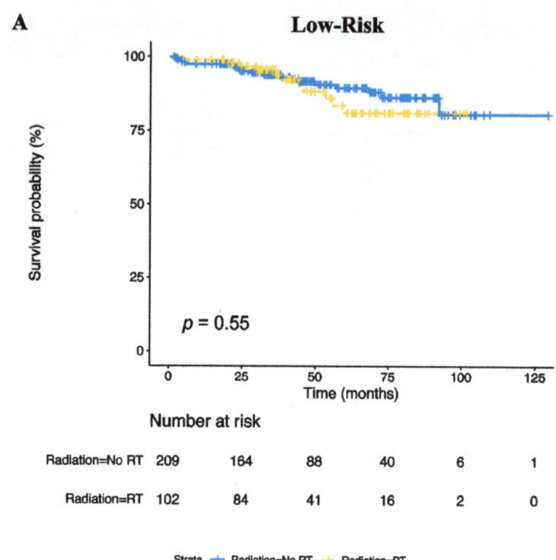

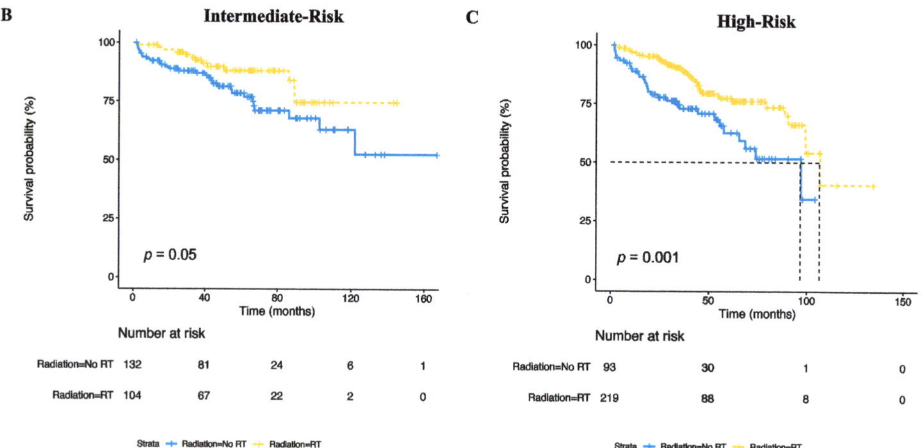

Figure 2. Kaplan–Meier curves for overall survival in the NCDB for low-risk (**A**), intermediate-risk (**B**), and high-risk (**C**) groups, based on the receipt of radiotherapy (RT).

Table 3. Summary of the association between radiotherapy and overall survival in NCDB, according to risk group.

Variable	Univariable			Multivariable		
	HR of Radiotherapy	95% CI	*p*-Value	HR of Radiotherapy	95% CI	*p*-Value
Risk Group						
Low risk	1.26	0.60, 2.65	0.55	-	-	
Intermediate risk	0.52	0.27, 0.99	**0.048**	0.74	0.38, 1.47	0.39
High risk	0.46	0.29, 0.74	**0.001**	0.59	0.36, 0.95	**0.031**

HR—hazard ratio, CI—confidence interval. The table provides summary statistics on the association of radiotherapy, with overall survival from separate univariable and multivariable analyses, stratified by risk group. Full univariable and multivariable analyses are included in the Supplementary Materials (Tables S4–S6). Variables included in the analysis were age, sex, race, Charlson–Deyo comorbidity index, tumor size, anatomical site, the extent of resection, and radiotherapy. Variables that were significant in the univariable analysis were included in the multivariable analyses. Bold values are statistically significant.

Because patients who died very early would not have had the opportunity to undergo radiotherapy, our outcome of interest may be affected by the immortal time bias [56]. To address the immortal time bias, we performed a sequential landmark time analysis. For patients who received radiotherapy, the time from diagnosis to the initiation of radiotherapy was available for 414 (97.4%) patients. The median time from diagnosis to the initiation of radiotherapy was 60 days, suggesting that our initial exclusion period of patients with less than 2 months of follow-up was appropriate. We performed additional landmark analyses for those patients with at least 3 and 6 months of follow-up. Similar results were observed in the high-risk group, with HRs of 0.51 (0.31–0.84, p = 0.008, Table S7 in the Supplementary Materials) and 0.54 (0.32–0.90 p = 0.02, Table S8 in the Supplementary Materials), respectively. In the intermediate-risk group, radiotherapy was no longer associated with a statistically significant improvement in OS in the 3-month (HR 0.57 (HR 0.29–1.11), p = 0.10, Table S9 in the Supplementary Materials) and 6-month analyses (HR 0.65 (0.32–1.31), p = 0.23, Table S10 in the Supplementary Materials).

4. Discussion

In this study, we develop and validate a risk-stratification schema for SFT/HPCs of the CNS according to the WHO 2016 histological guidelines, stratified by grade and EOR. Our risk categories were prognostic of OS and CSS and predicted outcomes better than any single clinical factor. Furthermore, our risk categories stratified patients to determine the survival benefit associated with radiotherapy. These risk categories may be used in future prospective trials or retrospective studies that evaluate the survival benefit of adjuvant radiotherapy.

The OS advantage observed with radiotherapy in the univariable analysis was limited to those patients with intermediate- or high-risk disease. Low-risk patients had a comparatively favorable prognosis and did not seem to experience a survival benefit from radiotherapy. Furthermore, at a median follow-up time of 80 months in the SEER dataset, there were no cause-specific deaths in the low-risk group. This suggests that radiotherapy can potentially be deferred in low-risk patients without affecting survival.

For high-risk tumors, the prognosis was poor, with most cause-specific deaths occurring in this group. Radiotherapy was associated with a reduction in mortality by over 50%, suggesting that radiotherapy is essential for disease management. This benefit remained robust on multivariable analysis and in multiple landmark sensitivity analyses. Given the poor prognosis of the disease, regardless of treatment, clinical trials that access treatment-escalation in this group beyond adjuvant radiotherapy may be appropriate.

In the intermediate-risk group, radiotherapy was associated with improved OS in the univariable analysis. However, it was no longer associated with improved OS in the multivariable analysis when including patient age, suggesting that the survival advantage may be confounded by patient selection. Additionally, the association was no longer statistically significant in our time-dependent landmark sensitivity analysis, which further supports the notion that patient selection is at least partially driving the observed effect. There were few cause-specific deaths in the intermediate-risk group in the SEER dataset at a median follow-up time of 74 months. Still, we cannot rule out the possibility that radiotherapy improves survival at later time points after 10 years. Even if radiotherapy does not improve OS, a progression-free survival benefit cannot be ruled out. Prolonging the time to progression may be associated with decreased morbidity and should be weighed against the potential toxicity from adjuvant radiotherapy. Toxicity datahave been reported in the RTOG and EORTC studies at dose levels of 54 and 60 Gy [30,31,34]. In the absence of randomized data, adjuvant radiotherapy should be considered at the clinical discretion of the treating provider, after a discussion of the risks and benefits.

Our study was inspired by the RTOG and EORTC trials, which successfully enrolled and completed prospective studies on risk-adapted radiotherapeutic strategies for meningiomas. These trials created established standard protocols for the treatment of meningioma and also led to two randomized phase-III trials in the US and Europe. The ongoing NRG-

BN003 and EORTC/ROAM trials will evaluate the role of adjuvant radiotherapy in grade-II meningiomas that undergo GTR. RTOG 0539 also demonstrated the feasibility of recruiting a high-risk group of patients with grade-III meningiomas, which are relatively rare [31]. Using a similar framework, we applied risk-stratification classes to SFT/HPC, based on prognostic groupings. Due to the rarity of SFT/HPC, a prospective study is unlikely. However, with an estimated 230 cases of SFT/HPC per year, versus 320 cases of malignant meningioma, a prospective trial may be feasible [1,2,57]. The last available estimate of the incidence rate is from 2013, and the incidence rate may have risen since then. Unlike extracranial SFT/HPC, the incidence rate of CNS SFT/HPC is slightly higher in Asian/Pacific Islanders [1,2]. Large series of CNS SFT/HPCs have been published from Asian countries and recruitment for trials may be more feasible in Asia.

Risk categories in our study were developed based on the overall survival prognosis and not on progression-free survival and varied from the RTOG 0539 study as follows: grade 2 SFT/HPC tumors with GTR were categorized as low-risk in our study, whereas grade 2 meningiomas that underwent GTR were considered intermediate-risk in the RTOG study; grade 3 SFT/HPCs with GTR were considered high-risk in the RTOG study, whereas they were classified as intermediate risk in the current study. Additionally, we only analyzed newly diagnosed tumors, whereas the RTOG 0539 study included recurrent tumors as well.

The advantages of our study include the use of two large national datasets. NCDB covers 70% of the US population and contains detailed treatment information, whereas SEER covers 28% of the US and is representative of the population. It also has cause-specific death information. Because we analyzed patients from 2004–2016 in the NCDB and 2000–2019 in SEER, we expected a considerable overlap of patients. These analyses were intended to be complementary, as data collection and quality control differ and because different variables are available.

The limitations of our study include retrospective analysis. Given the rarity of the tumor in question, there have been no prospective studies and and future prospective studiesare unlikely. A central histological review, including molecular analysis, was not possible. Although the EOR variable in the NCDB has been validated via data submitted from an academic center, the accuracy of EOR coding from nationwide samples is unknown [37]. Our analysis was corroborated in the SEER dataset, which may have better quality control procedures and less missing data. Radiotherapy may be under-coded in national datasets, which would bias our data toward the null hypothesis [1,17]. In the absence of prospective trials, large retrospective multi-institutional cohorts would be useful to validate our findings.

Our risk categories are pragmatic and may be applied in clinical scenarios when considering overall or cause-specific mortality for an individual patient. We advise that decisions for adjuvant treatment should be discussed within a multidisciplinary tumor board.

5. Conclusions

SFT/HPC of the CNS is a rare meningeal tumor, with no current consensus on the standard of care for adjuvant management. In this study, we develop and validate a risk-stratification schema based on the tumor grade and EOR, which is similar to risk classes developed for the RTOG 0539 and EORTC 22042-26042 trials. Our risk categories were prognostic of OS and CSS and outperformed the prognostic capability of any individual risk factor. Furthermore, our risk groups were predictive of survival benefits from radiotherapy. Radiotherapy was associated with an improved OS in the intermediate- and high-risk groups but not in the low-risk group. There were no cause-specific deaths in the low-risk group, suggesting that radiotherapy can be deferred without affecting survival. The OS benefit was not statistically significant in themultivariable analysis or in our sensitivity analyses in the intermediate-risk group, suggesting that the survival benefit may be, at least partially, driven by patient selection. Still, radiotherapy may be associated with a progression-free survival benefit and this may translate into an OS benefit at later follow-up times. In the high-risk group, radiotherapy was associated with reduced mortality, suggesting that it is essential for the management of grade 3 tumors. Prognosis is poor for grade 3 tumors, and investigation of additional therapy-escalation may be warranted.

These risk categories may be used as the basis for a prospective trial. Although a prospective study is unlikely, it may be feasible, given the rising incidence of SFT/HPC, the proof of feasibility already having been established when studying malignant meningioma in the RTOG 0539 trial. In the absence of prospective data, validation of our risk categories through a multi-institutional retrospective series would help in developing evidence-based management strategies for this rare tumor.

Supplementary Materials: The following supporting information can be downloaded at: https://www.mdpi.com/article/10.3390/cancers15030876/s1, Figure S1: Kaplan-Meier curves for overall survival in the NCDB (**A**) and overall (**B**) and cause-specific survival (**C**) in the SEER database based on grade and EOR. Figure S2: Sensitivity Analysis: Kaplan-Meier curves for overall survival in the NCDB, including patients only with all histological data points. Table S1: Demographic and clinical characteristics of the SEER cohort. Table S2: Univariable and multivariable analysis of overall survival in the NCDB. Table S3: Univariable and multivariable analysis of overall survival in SEER. Table S4: Univariable and multivariable analysis of overall survival in the low-risk group. Table S5: Univariable and multivariable analysis of overall survival in the intermediate-risk group. Table S6: Univariable and multivariable analysis of overall survival in the high-risk group. Table S7: Univariable and multivariable analysis of overall survival in the high-risk group at the 3-month landmark. Table S8: Univariable and multivariable analysis of overall survival in the high-risk group at the 6-month landmark. Table S9: Univariable and multivariable analysis of overall survival in the intermediate-risk group at the 3-month landmark. Table S10: Univariable and multivariable analysis of overall survival in the intermediate-risk group at the 6-month landmark.

Author Contributions: Conceptualization, C.J.K., S.K.C. and T.J.C.W.; methodology, C.J.K., S.K.C. and T.J.C.W.; formal analysis, C.J.K., P.K. and A.I.R.; investigation, C.J.K., P.K. and A.I.R.; resources, A.I.R., S.K.C. and T.J.C.W.; data curation, C.J.K. and A.I.R.; writing—original draft preparation, C.J.K.; writing—review and editing, C.J.K., A.I.R., P.K., G.M.M., M.B.S., J.N.B., J.B.Y., S.K.C. and T.J.C.W.; visualization, C.J.K., S.K.C. and T.J.C.W.; supervision, S.K.C. and T.J.C.W.; project administration, S.K.C. and T.J.C.W.; funding acquisition, S.K.C. and T.J.C.W. All authors have read and agreed to the published version of the manuscript.

Funding: This research received no external funding.

Institutional Review Board Statement: Not applicable.

Informed Consent Statement: Not applicable.

Data Availability Statement: The SEER and NCDB datasets are available upon request from the NCI and ACS. Coding scripts are available from the authors upon request.

Conflicts of Interest: No outside funding was received to support this work. Dr. Cheng reports receiving personal fees and nonfinancial support from AbbVie outside the submitted work. Dr. Wang reports personal fees and non-financial support from AbbVie, personal fees from Cancer Panels, personal fees from Doximity, personal fees and non-financial support from Elekta, personal fees and non-financial support from Merck, personal fees and non-financial support from Novocure, personal fees and non-financial support from RTOG Foundation, personal fees from Rutgers, personal fees from the University of Iowa, personal fees from Wolters Kluwer, grants and non-financial support from Genentech, grants and non-financial support from Varian, and personal fees from Iylon Precision Oncology, outside the submitted work.

References

1. Kinslow, C.J.; Bruce, S.S.; Rae, A.I.; Sheth, S.A.; McKhann, G.M.; Sisti, M.B.; Bruce, J.N.; Sonabend, A.M.; Wang, T.J.C. Solitary-fibrous tumor/hemangiopericytoma of the central nervous system: A population-based study. *J. Neuro-Oncol.* **2018**, *138*, 173–182. [CrossRef] [PubMed]
2. Kinslow, C.J.; Wang, T.J.C. Incidence of extrameningeal solitary fibrous tumors. *Cancer* **2020**, *126*, 4067. [CrossRef] [PubMed]

3. Louis, D.N.; Perry, A.; Reifenberger, G.; von Deimling, A.; Figarella-Branger, D.; Cavenee, W.K.; Ohgaki, H.; Wiestler, O.D.; Kleihues, P.; Ellison, D.W. The 2016 World Health Organization Classification of Tumors of the Central Nervous System: A summary. *Acta Neuropathol.* **2016**, *131*, 803–820. [CrossRef] [PubMed]

4. Bisceglia, M.; Galliani, C.; Giannatempo, G.; Lauriola, W.; Bianco, M.; D'Angelo, V.; Pizzolitto, S.; Vita, G.; Pasquinelli, G.; Magro, G.; et al. Solitary fibrous tumor of the central nervous system: A 15-year literature survey of 220 cases (August 1996–July 2011). *Adv. Anat. Pathol.* **2011**, *18*, 356–392. [CrossRef]

5. Bouvier, C.; Metellus, P.; de Paula, A.M.; Vasiljevic, A.; Jouvet, A.; Guyotat, J.; Mokhtari, K.; Varlet, P.; Dufour, H.; Figarella-Branger, D. Solitary fibrous tumors and hemangiopericytomas of the meninges: Overlapping pathological features and common prognostic factors suggest the same spectrum of tumors. *Brain Pathol. (Zur. Switz.)* **2012**, *22*, 511–521. [CrossRef] [PubMed]

6. Tihan, T.; Viglione, M.; Rosenblum, M.K.; Olivi, A.; Burger, P.C. Solitary fibrous tumors in the central nervous system. A clinicopathologic review of 18 cases and comparison to meningeal hemangiopericytomas. *Arch. Pathol. Lab. Med.* **2003**, *127*, 432–439. [CrossRef]

7. Mena, H.; Ribas, J.L.; Pezeshkpour, G.H.; Cowan, D.N.; Parisi, J.E. Hemangiopericytoma of the central nervous system: A review of 94 cases. *Hum. Pathol.* **1991**, *22*, 84–91. [CrossRef] [PubMed]

8. Guthrie, B.L.; Ebersold, M.J.; Scheithauer, B.W.; Shaw, E.G. Meningeal hemangiopericytoma: Histopathological features, treatment, and long-term follow-up of 44 cases. *Neurosurgery* **1989**, *25*, 514–522. [CrossRef]

9. Rutkowski, M.J.; Jian, B.J.; Bloch, O.; Chen, C.; Sughrue, M.E.; Tihan, T.; Barani, I.J.; Berger, M.S.; McDermott, M.W.; Parsa, A.T. Intracranial hemangiopericytoma: Clinical experience and treatment considerations in a modern series of 40 adult patients. *Cancer* **2012**, *118*, 1628–1636. [CrossRef]

10. Rutkowski, M.J.; Bloch, O.; Jian, B.J.; Chen, C.; Sughrue, M.E.; Tihan, T.; Barani, I.J.; Berger, M.S.; McDermott, M.W.; Parsa, A.T. Management of recurrent intracranial hemangiopericytoma. *J. Clin. Neurosci.* **2011**, *18*, 1500–1504. [CrossRef]

11. Ambrosini-Spaltro, A.; Eusebi, V. Meningeal hemangiopericytomas and hemangiopericytoma/solitary fibrous tumors of extracranial soft tissues: A comparison. *Virchows Arch. Int. J. Pathol.* **2010**, *456*, 343–354. [CrossRef]

12. Mathieu, D. Why do hemangiopericytomas have such high recurrence rates? *Expert Rev. Anticancer Ther.* **2016**, *16*, 1095–1096. [CrossRef]

13. Bastin, K.T.; Mehta, M.P. Meningeal hemangiopericytoma: Defining the role for radiation therapy. *J. Neurooncol.* **1992**, *14*, 277–287. [CrossRef] [PubMed]

14. Schiariti, M.; Goetz, P.; El-Maghraby, H.; Tailor, J.; Kitchen, N. Hemangiopericytoma: Long-term outcome revisited. *J. Neurosurg.* **2011**, *114*, 747–755. [CrossRef] [PubMed]

15. Kinslow, C.J.; Rajpara, R.S.; Wu, C.-C.; Bruce, S.S.; Canoll, P.D.; Wang, S.-H.; Sonabend, A.M.; Sheth, S.A.; McKhann, G.M.; Sisti, M.B.; et al. Invasiveness is associated with metastasis and decreased survival in hemangiopericytoma of the central nervous system. *J. Neuro-Oncol.* **2017**, *133*, 409–417. [CrossRef] [PubMed]

16. Louis, D.N.; Perry, A.; Wesseling, P.; Brat, D.J.; Cree, I.A.; Figarella-Branger, D.; Hawkins, C.; Ng, H.K.; Pfister, S.M.; Reifenberger, G.; et al. The 2021 WHO Classification of Tumors of the Central Nervous System: A summary. *Neuro. Oncol.* **2021**, *23*, 1231–1251. [CrossRef]

17. Noone, A.M.; Lund, J.L.; Mariotto, A.; Cronin, K.; McNeel, T.; Deapen, D.; Warren, J.L. Comparison of SEER Treatment Data With Medicare Claims. *Med. Care* **2016**, *54*, e55–e64. [CrossRef]

18. Ghose, A.; Guha, G.; Kundu, R.; Tew, J.; Chaudhary, R. CNS Hemangiopericytoma: A Systematic Review of 523 Patients. *Am. J. Clin. Oncol.* **2014**, *40*, 223–227. [CrossRef] [PubMed]

19. Sonabend, A.M.; Zacharia, B.E.; Goldstein, H.; Bruce, S.S.; Hershman, D.; Neugut, A.I.; Bruce, J.N. The role for adjuvant radiotherapy in the treatment of hemangiopericytoma: A Surveillance, Epidemiology, and End Results analysis. *J. Neurosurg.* **2014**, *120*, 300–308. [CrossRef]

20. Stessin, A.M.; Sison, C.; Nieto, J.; Raifu, M.; Li, B. The Role of Postoperative Radiation Therapy in the Treatment of Meningeal Hemangiopericytoma—Experience From the SEER Database. *Int. J. Radiat. Oncol.* **2013**, *85*, 784–790. [CrossRef]

21. Ghia, A.J.; Allen, P.K.; Mahajan, A.; Penas-Prado, M.; McCutcheon, I.E.; Brown, P.D. Intracranial hemangiopericytoma and the role of radiation therapy: A population based analysis. *Neurosurgery* **2013**, *72*, 203–209. [CrossRef] [PubMed]

22. Lee, E.J.; Kim, J.H.; Park, E.S.; Khang, S.K.; Cho, Y.H.; Hong, S.H.; Kim, C.J. The impact of postoperative radiation therapy on patterns of failure and survival improvement in patients with intracranial hemangiopericytoma. *J. Neuro-Oncol.* **2016**, *127*, 181–190. [CrossRef]

23. Chen, L.-F.; Yang, Y.; Yu, X.-G.; Gui, Q.-P.; Xu, B.-N.; Zhou, D.-B. Multimodal treatment and management strategies for intracranial hemangiopericytoma. *J. Clin. Neurosci.* **2015**, *22*, 718–725. [CrossRef] [PubMed]

24. Choi, J.; Park, S.-H.; Khang, S.K.; Suh, Y.-L.; Kim, S.P.; Lee, Y.S.; Kwon, H.S.; Kang, S.-G.; Kim, S.H. Hemangiopericytomas in the Central Nervous System: A Multicenter Study of Korean Cases with Validation of the Usage of STAT6 Immunohistochemistry for Diagnosis of Disease. *Ann. Surg. Oncol.* **2016**, *23*, 954–961. [CrossRef] [PubMed]

25. Zhang, G.J.; Wu, Z.; Zhang, L.W.; Li, D.; Zhang, J.T. Surgical management and adverse factors for recurrence and long-term survival in hemangiopericytoma patients. *World Neurosurg.* **2017**, *104*, 95–103. [CrossRef] [PubMed]

26. Rutkowski, M.J.; Sughrue, M.E.; Kane, A.J.; Aranda, D.; Mills, S.A.; Barani, I.J.; Parsa, A.T. Predictors of mortality following treatment of intracranial hemangiopericytoma. *J. Neurosurg.* **2010**, *113*, 333–339. [CrossRef]

27. Staples, J.J.; Robinson, R.A.; Wen, B.C.; Hussey, D.H. Hemangiopericytoma–the role of radiotherapy. *Int. J. Radiat. Oncol. Biol. Phys.* **1990**, *19*, 445–451. [CrossRef]
28. Ghia, A.J.; Chang, E.L.; Allen, P.K.; Mahajan, A.; Penas-Prado, M.; McCutcheon, I.E.; Brown, P.D. Intracranial hemangiopericytoma: Patterns of failure and the role of radiation therapy. *Neurosurgery* **2013**, *73*, 624–630. [CrossRef] [PubMed]
29. Kumar, A.; Guss, Z.D.; Courtney, P.T.; Nalawade, V.; Sheridan, P.; Sarkar, R.R.; Banegas, M.P.; Rose, B.S.; Xu, R.; Murphy, J.D. Evaluation of the Use of Cancer Registry Data for Comparative Effectiveness Research. *JAMA Netw. Open* **2020**, *3*, e2011985. [CrossRef]
30. Rogers, L.; Zhang, P.; Vogelbaum, M.A.; Perry, A.; Ashby, L.S.; Modi, J.M.; Alleman, A.M.; Galvin, J.; Brachman, D.; Jenrette, J.M.; et al. Intermediate-risk meningioma: Initial outcomes from NRG Oncology RTOG 0539. *J. Neurosurg. JNS* **2018**, *129*, 35–47. [CrossRef]
31. Rogers, C.L.; Won, M.; Vogelbaum, M.A.; Perry, A.; Ashby, L.S.; Modi, J.M.; Alleman, A.M.; Galvin, J.; Fogh, S.E.; Youssef, E.; et al. High-risk Meningioma: Initial Outcomes From NRG Oncology/RTOG 0539. *Int. J. Radiat. Oncol. Biol. Phys.* **2020**, *106*, 790–799. [CrossRef] [PubMed]
32. Rogers, C.L.; Pugh, S.L.; Vogelbaum, M.A.; Perry, A.; Ashby, L.S.; Modi, J.M.; Alleman, A.M.; Barani, I.J.; Braunstein, S.; Bovi, J.A.; et al. Low-risk meningioma: Initial outcomes from NRG Oncology/RTOG 0539. *Neuro Oncol.* **2022**, noac137. [CrossRef]
33. Rogers, C.L.; Perry, A.; Pugh, S.; Vogelbaum, M.A.; Brachman, D.; McMillan, W.; Jenrette, J.; Barani, I.; Shrieve, D.; Sloan, A.; et al. Pathology concordance levels for meningioma classification and grading in NRG Oncology RTOG Trial 0539. *Neuro Oncol.* **2015**, *18*, 565–574. [CrossRef] [PubMed]
34. Weber, D.C.; Ares, C.; Villa, S.; Peerdeman, S.M.; Renard, L.; Baumert, B.G.; Lucas, A.; Veninga, T.; Pica, A.; Jefferies, S.; et al. Adjuvant postoperative high-dose radiotherapy for atypical and malignant meningioma: A phase-II parallel non-randomized and observation study (EORTC 22042-26042). *Radiother. Oncol.* **2018**, *128*, 260–265. [CrossRef] [PubMed]
35. Bilimoria, K.Y.; Stewart, A.K.; Winchester, D.P.; Ko, C.Y. The National Cancer Data Base: A Powerful Initiative to Improve Cancer Care in the United States. *Ann. Surg. Oncol.* **2008**, *15*, 683–690. [CrossRef]
36. Davis, F.G.; McCarthy, B.J.; Berger, M.S. Centralized databases available for describing primary brain tumor incidence, survival, and treatment: Central Brain Tumor Registry of the United States; Surveillance, Epidemiology, and End Results; and National Cancer Data Base. *Neuro Oncol.* **1999**, *1*, 205–211. [CrossRef]
37. Harary, M.; Kavouridis, V.K.; Torre, M.; Zaidi, H.A.; Chukwueke, U.N.; Reardon, D.A.; Smith, T.R.; Iorgulescu, J.B. Predictors and early survival outcomes of maximal resection in WHO grade II 1p/19q-codeleted oligodendrogliomas. *Neuro Oncol.* **2020**, *22*, 369–380. [CrossRef]
38. Iorgulescu, J.B.; Torre, M.; Harary, M.; Smith, T.R.; Aizer, A.A.; Reardon, D.A.; Barnholtz-Sloan, J.S.; Perry, A. The Misclassification of Diffuse Gliomas: Rates and Outcomes. *Clin. Cancer Res.* **2019**, *25*, 2656–2663. [CrossRef]
39. Kinslow, C.J.; Canoll, P.; Cheng, S.K.; Wang, T.J.C. Misclassification of Diffuse Gliomas—Letter. *Clin. Cancer Res.* **2020**, *26*, 1198. [CrossRef]
40. Overview of the SEER Program. Available online: https://seer.cancer.gov/about/overview.html (accessed on 15 September 2020).
41. Kinslow, C.J.; Kim, A.; Sanchez, G.I.; Cheng, S.K.; Kachnic, L.A.; Neugut, A.I.; Horowitz, D.P. Incidence of Anaplastic Large-Cell Lymphoma of the Breast in the US, 2000 to 2018. *JAMA Oncol.* **2022**, *8*, 1354–1356. [CrossRef]
42. Kinslow, C.J.; DeStephano, D.M.; Rohde, C.H.; Kachnic, L.A.; Cheng, S.K.; Neugut, A.I.; Horowitz, D.P. Risk of Anaplastic Large Cell Lymphoma Following Postmastectomy Implant Reconstruction in Women With Breast Cancer and Ductal Carcinoma in Situ. *JAMA Netw. Open* **2022**, *5*, e2243396. [CrossRef]
43. Kinslow, C.J.; May, M.S.; Saqi, A.; Shu, C.A.; Chaudhary, K.R.; Wang, T.J.C.; Cheng, S.K. Large-Cell Neuroendocrine Carcinoma of the Lung: A Population-Based Study. *Clin. Lung Cancer* **2020**, *21*, e99–e113. [CrossRef] [PubMed]
44. May, M.S.; Kinslow, C.J.; Adams, C.; Saqi, A.; Shu, C.A.; Chaudhary, K.R.; Wang, T.J.C.; Cheng, S.K. Outcomes for localized treatment of large cell neuroendocrine carcinoma of the lung in the United States. *Transl. Lung. Cancer Res.* **2021**, *10*, 71–79. [CrossRef] [PubMed]
45. Zhou, Z.; Kinslow, C.J.; Hibshoosh, H.; Guo, H.; Cheng, S.K.; He, C.; Gentry, M.S.; Sun, R.C. Clinical Features, Survival and Prognostic Factors of Glycogen-Rich Clear Cell Carcinoma (GRCC) of the Breast in the U.S. Population. *J. Clin. Med.* **2019**, *8*, 246. [CrossRef]
46. Zhou, Z.; Kinslow, C.J.; Wang, P.; Huang, B.; Cheng, S.K.; Deutsch, I.; Gentry, M.S.; Sun, R.C. Clear Cell Adenocarcinoma of the Urinary Bladder Is a Glycogen-Rich Tumor with Poorer Prognosis. *J. Clin. Med.* **2020**, *9*, 138. [CrossRef] [PubMed]
47. Kinslow, C.J.; May, M.S.; Kozak, M.; Pollom, E.L.; Chang, D.T. Signet ring cell carcinoma of the Ampulla of Vater: Outcomes of patients in the United States. *HPB* **2020**, *22*, 1759–1765. [CrossRef]
48. Facility Oncology Registry Data Standards (FORDS): Revised for 2013. Available online: https://www.facs.org/~{}/media/files/quality%20programs/cancer/coc/fords/fords%20manual%202013.ashx (accessed on 15 December 2022).
49. Garton, A.L.A.; Kinslow, C.J.; Rae, A.I.; Mehta, A.; Pannullo, S.C.; Magge, R.S.; Ramakrishna, R.; McKhann, G.M.; Sisti, M.B.; Bruce, J.N.; et al. Extent of resection, molecular signature, and survival in 1p19q-codeleted gliomas. *J. Neurosurg. JNS* **2021**, *134*, 1357–1367. [CrossRef]
50. Kinslow, C.J.; Garton, A.L.A.; Rae, A.I.; Marcus, L.P.; Adams, C.M.; McKhann, G.M.; Sisti, M.B.; Connolly, E.S.; Bruce, J.N.; Neugut, A.I.; et al. Extent of resection and survival for oligodendroglioma: A U.S. population-based study. *J. Neuro-Oncol.* **2019**, *144*, 591–601. [CrossRef]

51. Rae, A.I.; Mehta, A.; Cloney, M.; Kinslow, C.J.; Wang, T.J.C.; Bhagat, G.; Canoll, P.D.; Zanazzi, G.J.; Sisti, M.B.; Sheth, S.A.; et al. Craniotomy and Survival for Primary Central Nervous System Lymphoma. *Neurosurgery* **2019**, *84*, 935–944. [CrossRef]
52. Kinslow, C.J.; Rae, A.I.; Neugut, A.I.; Adams, C.M.; Cheng, S.K.; Sheth, S.A.; McKhann, G.M.; Sisti, M.B.; Bruce, J.N.; Iwamoto, F.M.; et al. Surgery plus adjuvant radiotherapy for primary central nervous system lymphoma. *Br. J. Neurosurg.* **2020**, *34*, 690–696. [CrossRef]
53. Boyett, D.; Kinslow, C.J.; Bruce, S.S.; Sonabend, A.M.; Rae, A.I.; McKhann, G.M.; Sisti, M.B.; Bruce, J.N.; Cheng, S.K.; Wang, T.J.C. Spinal location is prognostic of survival for solitary-fibrous tumor/hemangiopericytoma of the central nervous system. *J. Neuro-Oncol.* **2019**, *143*, 457–464. [CrossRef]
54. Surveillance, Epidemiology, and End Results (SEER) Program. SEER*Stat Database: Incidence—SEER Research Data, 17 Registries, Nov 2020 Sub (1975–2019)—Linked To County Attributes—Time Dependent (1990–2019) Income/Rurality, 1969–2019 Counties, National Cancer Institute, DCCPS, Surveillance Research Program, Released 21, Based on the November 2020 Submission. Available online: www.seer.cancer.gov (accessed on 15 December 2022).
55. Vasista, A.; Stockler, M.R.; West, T.; Wilcken, N.; Kiely, B.E. More than just the median: Calculating survival times for patients with HER2 positive, metastatic breast cancer using data from recent randomised trials. *Breast* **2017**, *31*, 99–104. [CrossRef] [PubMed]
56. Newman, N.B.; Brett, C.L.; Kluwe, C.A.; Patel, C.G.; Attia, A.; Osmundson, E.C.; Kachnic, L.A. Immortal Time Bias in National Cancer Database Studies. *Int. J. Radiat. Oncol. Biol. Phys.* **2020**, *106*, 5–12. [CrossRef] [PubMed]
57. Ostrom, Q.T.; Price, M.; Neff, C.; Cioffi, G.; Waite, K.A.; Kruchko, C.; Barnholtz-Sloan, J.S. CBTRUS Statistical Report: Primary Brain and Other Central Nervous System Tumors Diagnosed in the United States in 2015–2019. *Neuro Oncol.* **2022**, *24*, v1–v95. [CrossRef] [PubMed]

Article

Intracranial Solitary Fibrous Tumour Management: A French Multicentre Retrospective Study

Marine Lottin [1,*], Alexandre Escande [2,3,4], Luc Bauchet [5,6,7], Marie Albert-Thananayagam [8], Maël Barthoulot [8], Matthieu Peyre [9], Mathieu Boone [1], Sonia Zouaoui [5,7], Jacques Guyotat [10], Guillaume Penchet [11], Johan Pallud [12], Henry Dufour [13], Evelyne Emery [14], Michel Lefranc [15], Sébastien Freppel [16], Houman Namaki [17], Edouard Gueye [18], Jean-Jacques Lemaire [19], Bertrand Muckensturm [20], Robin Srour [21], Stéphane Derrey [22], Apolline Monfilliette [23], Jean-Marc Constans [24], Claude-Alain Maurage [25], Bruno Chauffert [26] and Nicolas Penel [8,27]

1. Department of Oncology, Amiens University Hospital, 80054 Amiens, France
2. Department of Radiotherapy, Lille Oscar Lambret Centre, 59000 Lille, France
3. School of Medicine H. Warembourg, Lille University, 59000 Lille, France
4. Lab CRIStAL, Lille University, UMR 9189, 59655 Villeneuve d'Ascq, France
5. Department of Neurosurgery, Montpellier University Hospital, 34295 Montpellier, France
6. Institute of Functional Genomics, Montpellier University, 34000 Montpellier, France
7. Unit of French Brain Tumour DataBase, (Registre des Tumeurs de l'Hérault), Montpellier Cancer Insitute, 34000 Montpellier, France
8. Clinical Research Department, Oscar Lambret Centre, 59000 Lille, France
9. Department of Neurosurgery, Sorbonne University, La-Pitié-Salpêtrière University Hospital, 75651 Paris, France
10. Department of Neurosurgery, Lyon University Hospital, 69500 Bron, France
11. Department of Neurosurgery, Bordeaux University Hospital, 33076 Bordeaux, France
12. Department of Neurosurgery, Sainte-Anne University Hospital, 75014 Paris, France
13. Department of Neurosurgery, La Timone University Hospital, 13005 Marseille, France
14. Department of Neurosurgery, Caen University Hospital, 14033 Caen, France
15. Department of Neurosurgery, Amiens University Hospital, 80054 Amiens, France
16. Department of Neurosurgery, Saint-Pierre University Hospital, 97448 Saint Pierre, France
17. Department of Neurosurgery, Perpignan Hospital, 66046 Perpignan, France
18. Department of Neurosurgery, Limoges University Hospital, 87042 Limoges, France
19. Department of Neurosurgery, Clermont-Ferrand University Hospital, 63003 Clermont-Ferrand, France
20. Department of Neurosurgery, Orleans Hospital, 45067 Orleans, France
21. Department of Neurosurgery, Colmar Hospital, 68024 Colmar, France
22. Department of Neurosurgery, Rouen University Hospital, 76038 Rouen, France
23. Department of Medical Oncology, Lille University Hospital, 59000 Lille, France
24. Department of Radiology, Amiens University Hospital, 80054 Amiens, France
25. Department of Pathology, Lille University Hospital, 59000 Lille, France
26. Department of Oncology, Saint Quentin Hospital, 02321 Saint Quentin, France
27. Department of Medical oncology, Oscar Lambret Centre, 59000 Lille, France
* Correspondence: lottin.marine@chu-amiens.fr; Tel.: +33-322455499

Citation: Lottin, M.; Escande, A.; Bauchet, L.; Albert-Thananayagam, M.; Barthoulot, M.; Peyre, M.; Boone, M.; Zouaoui, S.; Guyotat, J.; Penchet, G.; et al. Intracranial Solitary Fibrous Tumour Management: A French Multicentre Retrospective Study. *Cancers* **2023**, *15*, 704. https://doi.org/10.3390/cancers15030704

Academic Editor: Pierre Giglio

Received: 1 December 2022
Revised: 9 January 2023
Accepted: 20 January 2023
Published: 24 January 2023

Simple Summary: Intracranial solitary fibrous tumours (iSFTs) are exceptional mesenchymal tumours with a high relapse rate. We aimed to analyse the clinical outcome at each stage of the disease. We carried out a multicentre retrospective study including 88 patients from 16 French centres. Gross tumour resection was found to be a factor for good prognosis and significantly reduced local recurrence without impacting overall survival. High-grade tumours were a factor for poorer PFS and LRFS. More than 40% of our patients experienced local recurrence and were mostly treated with surgery and radiotherapy. The first relapse is a turning point in iSFT evolution, with reduced recurrence latency over the course of the disease. The management of repeated recurrence and disseminated diseases is challenging; these situations should be treated, if feasible, with local techniques considering the poor efficiency of systemic treatments.

Abstract: Background: Intracranial solitary fibrous tumour (iSFT) is an exceptional mesenchymal tumour with high recurrence rates. We aimed to analyse the clinical outcomes of newly diagnosed and recurrent iSFTs. Methods: We carried out a French retrospective multicentre ($n = 16$) study of histologically proven iSFT cases. Univariate and multivariate Cox models were used to estimate the prognosis value of the age, location, size, WHO grade, and surgical extent on overall survival (OS), progression-free survival (PFS), and local recurrence-free survival (LRFS). Results: Eighty-eight patients were included with a median age of 54.5 years. New iSFT cases were treated with gross tumour resection (GTR) ($n = 75$) or subtotal resection (STR) ($n = 9$) and postoperative radiotherapy (PORT) ($n = 32$, 57%). The median follow-up time was 7 years. The median OS, PFS, and LRFS were 13 years, 7 years, and 7 years, respectively. Forty-two patients experienced recurrence. Extracranial metastasis occurred in 16 patients. Median OS and PFS after the first recurrence were 6 years and 15.4 months, respectively. A higher histological grade was a prognosis factor for PFS ($p = 0.04$) and LRFS ($p = 0.03$). GTR influenced LRFS ($p = 0.03$). Conclusion: GTR provided benefits as a first treatment for iSFTs. However, approximately 40% of patients experienced relapse, which remains a challenging state.

Keywords: solitary fibrous tumour; hemangiopericytoma; intracranial; surgery; recurrence

1. Introduction

Intracranial solitary fibrous tumours (iSFTs) and hemangiopericytoma are rare primary tumours of the central nervous system (CNS) [1]. These mesenchymal tumours are considered sarcomas and represent less than 1% of primary CNS tumours and 2–5% of all meningeal tumours [1,2]. Their incidence in France is 0.061 per 100,000 inhabitants per year and 0.041 per 100,000 inhabitants worldwide [3]. ISFTs emerge from pericytes, a mural component of vessels that enables the function of the blood–brain barrier and the support of intracranial immunity [1,2]. Until recently, SFTs were distinguished from hemangiopericytoma (HPC) because of their difference in terms of aggressiveness. Even if SFTs were mostly considered benign tumours, some have a malignant evolution [1]. Conversely, HPC evolution may mimic benign tumours [1]. Thus, based on those characteristics and a common biological anomaly, i.e., the fusion of the NAB2-STAT6 gene arising from the chromosome 12q13 inversion, the 2016 WHO classification merged SFTs and HPC into the single entity "SFT/HPC" [4,5]. To align with the soft tissue nomenclature, this entity was renamed "SFT" in 2021.

Recurrences of iSFT are frequent, iterative, and more commonly located at the initial tumour site [6,7]. Because of their low occurrence, evidence for the general management of iSFTs still relies on few prospective data and retrospective studies [1,6,8,9]. Relapses are even less described, leading to heterogeneous practice between and within countries. Those relapses might also be metastatic with poor outcomes [10,11]. The aim of iSFT management must be to reduce the percentage of local and distant recurrences while considering practicable therapy [1,12]. Surgery is the universal mainstay of SFT treatment. A complete surgery or "gross tumour resection" (GTR) enables a significant gain in terms of survival and local control [1,2]. In addition, postoperative radiotherapy (PORT) provides good local control, especially after a "subtotal tumour resection" (STR) and high-grade cases. However, PORT's impact on the overall survival (OS) of iSFTs is still discussed, leading to various approaches to management at international and national levels [1,10,13,14]. Given the difficulty of performing successive local interventions without considerable side effects, systemic treatments might be employed to treat local recurrences. Systemic treatments are mostly used in the metastatic stage, although the impact of survival remains inconclusive [8,11]. Thus, the treatment of local and distant relapses remains a serious source of concern, lacking sufficient guidelines.

Currently, very few studies have examined the survival outcome of iSFTs according to their treatment following diagnosis and relapse [9,15,16]. Firstly, we aimed to analyse iSFT outcomes at each stage of the disease in a large, multicentre, retrospective French cohort.

Secondly, we aimed to analyse the associated clinicopathological factors to provide a better understanding and enhance future therapeutic decisions.

2. Materials and Methods

2.1. Case Selection

We collected the data of cases from the "French brain tumour database" (FBTDB) in collaboration with the French neuropathology network (RENOCLIP-LOC). The database combines information of patients with a confirmed pathological diagnosis of intracranial SFT, HPC, and anaplastic HPC in France from 2006 to 2015. Clinical and pathological records of the patients were retrospectively reviewed to extract the relevant clinical factors at 16 French centres (Amiens, Bordeaux, Caen, Clermont-Ferrand, Colmar/Strasbourg, Ile de la Réunion, Lille, Limoges, Lyon, Marseille, Montpellier, Orléans, Paris-La-pitié-salpêtrière, Paris-Sainte-Anne, Perpignan, and Rouen) from January 2006 to December 2015. The exclusion criteria were disease occurrence outside the inclusion period or at an extracranial level, a lack of histological proof of SFT, age <18 years at diagnosis, and a written refusal of consent to the study. Information on age, sex, clinical presentation of the disease, time to diagnosis, tumour location, tumour size, extent of resection, preoperative biopsy, preoperative embolization, pathology including immunohistochemistry, adjuvant radiotherapy, tumour recurrence or metastasis occurrence, and patient survival was collected.

Intracranial hypertension was defined by the association of headaches, nausea, and vomiting, and was distinguished from exclusive headaches in the clinical presentation of the disease. Tumour size corresponded to the maximal diameter measured by the radiologist on the preoperative radiologic sequences. Time to diagnosis was estimated as the period between the outbreak of evocative neurological symptoms and the date of the established histology. The extent of resection was classified as GTR or STR according to the postoperative MRI results. As meningiomas, GTR and STR are defined as Simpson grades I or II removal and grades III or IV removal, respectively [1,12]. Preoperative intervention through embolization guided with arteriography was conducted at the surgeon's discretion. Tumour local recurrence was defined as the reappearance of the tumour within the cranial cavity or an increase in the size of the residual tumour according to RECIST1.1 criteria. Metastasis was defined as an extracranial appearance of SFT.

2.2. Pathology

All specimens of SFTs were confirmed by a pathologist with expertise in this disease. Specimens were graded according to the 2021 CNS WHO classification into grades 1, 2, or 3 [5]. According to these criteria, grade 1 is characterized by a highly collagenous, relatively low-cellularity spindle-cell lesion, previously diagnosed as SFT. Grade 2 corresponds to a more cellular, less collagenous tumour with plump cells, staghorn vasculature, and mitosis <5 per 10 high-power field (HPF), previously diagnosed as HPC. Grade 3 corresponds to ≥ 5 mitoses/10 HPF and/or the presence of necrosis, and it was previously named "anaplastic HPC".

2.3. Statistical Analyses

Patient characteristics were described using the mean (standard deviation (SD)) or median (range) as appropriate for continuous variables and frequency (percentage) for categorical variables. Missing data were reported.

The median follow-up time was estimated by using the reverse Kaplan–Meier method, considering survival status at the end of the study as an event and death as a censored event. Survival analyses were performed using the Kaplan–Meier method. The survival rate, median, and confidence interval at 95% (95% confidence interval (95% CI)) were reported. Follow-up times were defined as the time from the date of the initial pathological diagnosis of SFT to the date of (1) death for OS; (2) recurrence or death if no recurrence occurred for progression-free survival (PFS); (3) local recurrence or death if no local recurrence occurred for local recurrence-free survival (LRFS). In cases of recurrence, follow-up times were defined as the time from the date of first recurrence of SFT to the date of (4) death for the second OS (OS2) and (5) the second recurrence or death if no second recurrence occurred for the second PFS (PFS2). Patients who remained alive or were lost-to-follow-up were censored.

The hazard ratio (HR) for recurrence and/or death (PFS) or for death (OS) associated with patients' characteristics were estimated using Cox models—first, in the univariate analysis and then in the multivariate analysis adjusted for possible confounders. The initial multivariate model included all covariates with a *p*-value < 0.20 in univariate analyses except for adjuvant radiotherapy, which only concerned the subgroup of patients with the indication according to the literature [14]. Correlations between variables were searched. The backward selection gave the final multivariate model including only covariates with *p*-value < 0.05. Sensitivity multivariate analyses were performed with PORT as a covariate. All the point estimates were reported with their 95% CI. All tests were two-sided, and the threshold for statistical significance was set to $p < 0.05$.

The software used for the analyses was Stata version 17.0 (StataCorp. 2021. Stata Statistical Software: Release 17. College Station, TX: StataCorp LLC).

2.4. Ethical Approval

The study complied with the "reference methodology" MR004 adopted by the French Data Protection Authority (CNIL), and we checked that patients did not object to the use of their clinical data for the research purpose. Ethical local approval was obtained on 25 May 2022. The number of the ethical approval is N° 2022-013. All data were anonymized.

3. Results

3.1. Patient Characteristics

A query of the database revealed 179 patients with a primary iSFT from January 2006 to December 2015 in the 16 participating French centres. Ninety-one cases were excluded because they were diagnosed before 2006 or after 2015 ($n = 37$), had extracranial localization ($n = 19$), were missing proof of the disease ($n = 28$), had a revised histology by pathological review ($n = 6$), or were patients aged <18 years old ($n = 1$). Finally, we included 88 patients in the study (Figure 1).

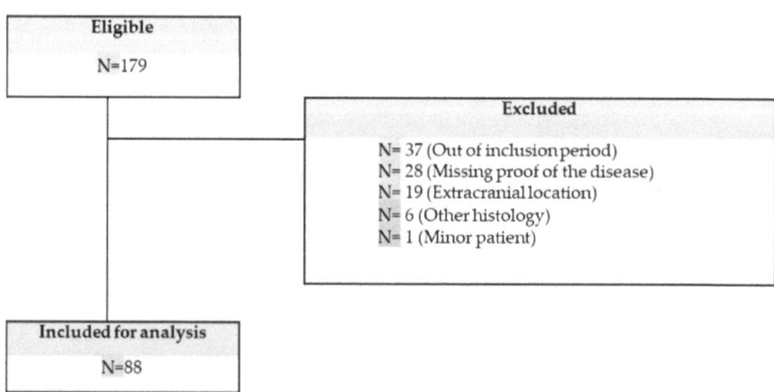

Figure 1. Flow chart of the study.

There was a female predominance ($n = 50$, 57%), and the median age at diagnosis was 54.5 years (range: 19–88). The SFTs were mainly local ($n = 84$, 95%) and supratentorial ($n = 71$, 81%) at diagnosis. Tumours often had a large maximal diameter on brain MRI ($\geq$5 cm for 32 (36%) patients-Table 1). All the clinical characteristics of the patients were similar regardless of tumour lateralization and are summarized in Table 1.

Table 1. Clinical, pathological, morphological, and treatment characteristics for the 88 patients presenting a new diagnosis of iSFT.

Characteristics (*N* = 88)	*n* (%)
Sex	
Female	50 (57%)
Male	38 (43%)
Median age (range)	54.5 (19–88)
Mean age (SD)	51.5 (16.4)
STAT6 expression (immunohistochemistry)	
Yes	34 (39%)
No	0 (0%)
Not available	54 (61%)
The WHO histoprognostic grade	
Grade 1	1 (1%)
Grade 2	25 (28%)
Grade 3	51 (58%)
Not available	11 (13%)
Mitotic count	
<5	36 (41%)
≥5	45 (51%)
Not available	7 (8%)
Necrosis	
Yes	33 (38%)
No	55 (63%)
Number of lesions	
Unique	84 (95%)
Multiple	4 (5%)
Maximal diameter	
< 3cm	5 (6%)
3–5 cm	24 (27%)
≥5 cm	32 (36%)
Not available	27 (30%)
Topography of the tumour	
Right-sided	42 (48%)
Left-sided	40 (45%)
Midline *	4 (4%)
Bilateral	1 (1%)
Not available	1 (1%)
Location	
Supratentorial	71 (81%)
Infratentorial	17 (17%)
Supra- and infratentorial	2 (2%)
Preoperative embolization	
Yes	8 (9%)
No	80 (91%)
Resection grade according to Simpson	
Simpson 1–2 (gross tumour resection)	75 (85%)
Simpson 3–4 (subtotal tumour resection)	9 (10%)
Simpson 5 (biopsy)	1 (1%)
Not available	3 (3%)

Table 1. *Cont.*

Characteristics (*N* = 88)	*n* (%)
Postoperative radiotherapy (*N* = 56)	
Yes	32 (57%)
No	24 (43%)
Technique of radiotherapy	
Conformational radiotherapy	19 (50%)
Radiosurgery	3 (9%)
Not available	10 (31%)
Local control	
Yes	76 (86%)
No	11 (13%)
Not available	1 (1%)
Type of treatment (*N* = 65)	
Surgery alone	33 (51%)
Radiotherapy alone	1 (1%)
Surgery plus postoperative radiotherapy	31 (48%)

* Falx and pineal region tumours.

3.2. Clinical and Pathological Presentation

The symptoms leading to diagnosis were related to the tumour location and consisted mainly of headaches (*n* = 41, 47%) and cognitive disorders (*n* = 37, 42%) (see Supplementary Table S1).

Seventy-seven tumours (70%) were graded according to the 2021 CNS WHO classification. There were 1 (1%), 25 (28%), and 51 (58%) grade one, two, and three iSFT cases, respectively. The 34 samples (39%) that were tested for the nuclear expression of STAT6 using immunohistochemistry were all positive (Table 1). The characteristics of the population with STAT6 status results were similar to those with an unknown status.

3.3. Initial Treatment

Eighty-seven (99%) patients underwent surgical resection and one (1%) had a biopsy without an additional resection. GTR and STR were achieved in 75 (85%) and 9 (10%) patients, respectively. The surgical excision (GTR or STR) status of the three other patients was unknown (Table 1). Only eight patients (9%) received preoperative embolization. After initial surgery, 32 patients (57%) received adjuvant radiotherapy. One of the 32 patients received exclusive radiotherapy after a single biopsy without resection. Systemic treatments were not employed at the initial stage (Table 1).

3.4. Survival Analyses

The median follow-up time was 7 years (range 0–16 years). Twenty-four patients had died by the retrospective analysis time, seven (29%) of whom had metastatic disease. The median OS time was 13 years (95% CI: 10–nonreached (NR)), with 1-, 5-, and 10-year OS rates of 93% (95% CI: 85–97), 85% (95% CI: 74–91), and 64% (95% CI: 48–76), respectively. The median PFS was 7 years (95% CI: 6–8 years) after the date of diagnosis, with 1-, 5-, and 10-year PFS rates of 87% (95% CI: 78–93), 66% (95% CI: 54–76), and 19% (95% CI: 9–32), respectively. The median LRFS was 7 years (95% CI: 6–8 years) after the date of diagnosis, with 1-, 5-, and 10-year LRFS rates of 87% (95% CI: 78–93), 68% (95% CI: 56–77), and 23% (95% CI: 11–36), respectively (Figure 2).

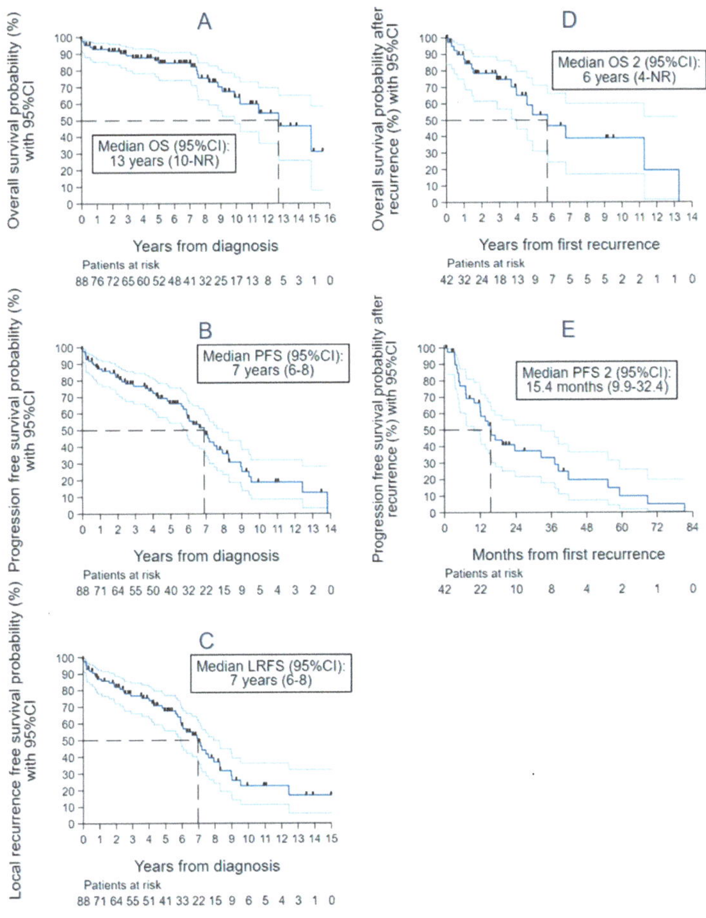

Figure 2. Survival analysis in newly diagnosed and recurrent iSFT patients. (**A**) Survival curve according to the Kaplan–Meier method showing median OS, (**B**) PFS, and (**C**) LRFS in newly diagnosed iSFTs and (**D**) OS2 and (**E**) PFS2 after the first recurrence, with their respective 95% confidence intervals (95% CI). NR: nonreached.

The median OS2 was 6 years (95% CI 4–NR), with 1-, 3-, and 5-year OS2 of 87% (95% CI 72–94), 75% (95% CI 57–86), and 53% (95% CI 31–71), respectively. The median PFS2 was 15.4 months (95% CI 9.9–32.4), with 1-, 3-, and 5-year PFS2 rates of 61% (95% CI 44–74), 33% (95% CI 18–49), and 10% (95% CI 2–26), respectively (Figure 2D,E).

3.5. Recurrences and Treatment at Recurrence

During the follow-up, 42 patients (48%) experienced at least one progression or recurrence and 16 (18%) patients had a distant recurrence. Among the 42 relapses, 35 (83%) were located at the initial site, 5 (12%) were distant, and 2 (5%) were simultaneously local and distant. The morphological and treatment characteristics of the first local recurrence are summarized in Table 2. Among the 37 patients presenting a local relapse, 17 (46%) received surgical treatment, of whom 10 (59%) underwent PORT. Exclusive radiotherapy was performed in 14 patients (38%) and systemic treatment in 2 (5%) (Table 2).

Table 2. Morphological and therapeutic management data in patients with at least one episode of localized recurrence of iSFT.

Characteristics ($N = 37$)	n (%)
Unifocal recurrence	
Yes	32 (87%)
No	5 (14%)
Maximal diameter of recurrence	
<3 cm	19 (51%)
3–5 cm	5 (14%)
≥5 cm	2 (5%)
Not available	11 (30%)
Recurrence topography	
Right-sided	18 (49%)
Left-sided	16 (43%)
Midline	3 (8%)
Bilateral	0 (0%)
Recurrence location	
Supratentorial	30 (81%)
Infratentorial	7 (19%)
Supra- and infratentorial	0 (0%)
Preoperative embolization	
Yes	0 (0%)
No	37 (100%)
Preoperative radiotherapy	
Yes	1 (3%)
No	36 (97%)
Surgery	
Yes	17 (46%)
No	20 (54%)
Resection grade according to Simpson ($N = 17$)	
Simpson 1–2 (gross tumour resection)	15 (88%)
Simpson 3–4 (subtotal tumour resection)	1 (6%)
Not available	1 (6%)
Postoperative radiotherapy ($N = 17$)	
Yes	10 (59%)
No	7 (41%)
Type of radiotherapy	
Conformational radiotherapy	4 (40%)
Radiosurgery	4 (40%)
Not available	2 (20%)
Exclusive radiotherapy	
Yes	14 (38%)
No	23 (62%)
Technique of radiotherapy	
Conformational radiotherapy	2 (14%)
Radiosurgery	8 (57%)
Not available	4 (29%)
Chemotherapy associated with radiotherapy	
Yes	1 (3%)
No	36 (97%)

Table 2. *Cont.*

Characteristics (*N* = 37)	*n* (%)
Postoperative chemotherapy	
Yes	1 (3%)
No	36 (97%)
Local control after treatment	
Yes	26 (70%)
No	10 (27%)
Not available	1 (3%)
Type of treatment that led to control (*N* = 26)	
Surgery alone	4 (15%)
Radiotherapy alone	11 (42%)
Surgery and postoperative radiotherapy	11 (42%)
Chemotherapy alone	0 (0%)
Radiotherapy and postoperative chemotherapy	0 (0%)
No treatment	0 (0%)

A second recurrence occurred in 25 patients, of whom 16 were local (64%) and treatment corresponded to surgery for 6 (38%) patients, radiotherapy for 6 (38%), and systemic treatment for 3 (19%) patients. The relapses were repetitive, and we observed fifth (*n* = 3), sixth (*n* = 2), and seventh (*n* = 1) recurrences in our series, which were all metastatic. Their treatment consisted of either surgery, radiotherapy, or systemic treatments. To our knowledge, no other solid or haematological cancer occurred during the time of surveillance.

3.6. Extracranial Metastasis

Sixteen patients (18%) developed extracranial metastases. The most frequent sites of extracranial metastases were the liver (five patients, 31%), bones (nine patients, 56%), and the lungs (six patients, 38%). Interestingly, only eight (50%) patients received a systemic treatment at this disseminated stage. The treatment consisted of chemotherapy (doxorubicin *n* = 2; temozolomide *n* = 2; doxorubicin + ifosfamide *n* = 1) in five (62.5%) patients and antiangiogenic targeted therapy in two patients (25%) (bevacizumab *n* = 1, pazopanib *n* = 1). One patient (12.5%) was treated with a combination of chemotherapy and vascular endothelial growth factor (VEGF) inhibitor (temozolomide and bevacizumab). A complete response was observed in five patients who underwent surgery or radiotherapy. The best response in the systemic group was a stable disease (see Supplementary Table S2).

3.7. Prognosis Factors

We observed that age was associated with poorer OS (HR = 1.04 (95% CI: 1.01–1.08); *p* = 0.01), PFS (HR = 1.03 (95% CI: 1.01–1.05); *p* = 0.008), and LRFS (HR = 1.04 (95% CI: 1.01–1.06); *p* = 0.002) (Table 3). Tumour topography (midline and left lateralization) was associated with worse OS in the multivariate analyses with HR = 2.69 (95% CI: 1.07–6.81), *p* = 0.03 for left tumours and HR = 8.29 (95% CI: 1.42–48.50), *p* = 0.03 for midline tumours, as compared with right-sided tumours (Table 3, Figure 3A). A higher histological grade was significantly associated with lower PFS and LRFS (HR = 2.14 (95% CI: 1.03–4.78; *p* = 0.04) and HR = 2.36 (95% CI: 1.08–5.16; *p* = 0.03), respectively) (Table 3).

Table 3. Univariate and multivariate analyses according to Cox models in patients presenting an initial diagnosis of iSFT.

Characteristics	Overall Survival				Progression-Free Survival				Local Recurrence-Free Survival			
	Univariate HR (95% CI)	*p*	Multivariate HR (95% CI)	*P*	Univariate HR (95% CI)	*p*	Multivariate HR (95% CI)	*P*	Univariate HR (95% CI)	*p*	Multivariate HR (95% CI)	*p*
Sex		0.78				0.36				0.60		
Female	1				1				1			
Male	0.89(0.39–2.02)				0.76 (0.43–1.36)				0.85(0.48–1.53)			
Age *	**1.05(1.02–1.09)**	**0.003**	**1.04(1.01–1.08)**	**0.01**	**1.03 (1.01–1.05)**	**0.003**	**1.03(1.01–1.05)**	**0.008**	**1.03(1.01–1.05)**	**0.002**	**1.04(1.01–1.06)**	**0.002**
Signs of raised intracranial pressure $^\$$ (*n* = 85)	1.30(0.53–3.17)	0.57			0.95(0.48–1.89)	0.89			0.99(0.50–1.97)	0.98		
Motor deficit $^\$$ (*n* = 85)	1.56 (0.63–3.84)	0.34			1.34 (0.71–2.51)	0.37			1.28 (0.67–2.44)	0.46		
Sensory deficit $^\$$ (*n* = 85)	1.11 (0.26–4.80)	0.89			0.90 (0.32–2.51)	0.84			0.95 (0.34–2.66)	0.92		
Epileptic seizures *$^\$$ (*n* = 85)	0.22 (0.03–1.66)	0.14			0.37 (0.13–1.03)	0.056			0.40 (0.14–1.11)	0.08		
High-function disorder *$^\$$ (*n* = 85)	1.65 (0.72–3.78)	0.23			1.52 (0.86–2.70)	0.15			1.52 (0.85–2.73)	0.16		
Visual disorder $^\$$ (*n* = 85)	0.60 (0.20–1.77)	0.35			0.95 (0.50–1.81)	0.88			1.02 (0.53–1.95)	0.95		
Cerebellar syndrome *$^\$$ (*n* = 86)	1.83 (0.67–5.00)	0.24			1.78 (0.83–3.86)	0.14			1.86 (0.86–4.02)	0.12		
Headache $^\$$ (*n* = 85)	1.00 (0.43–2.33)	1.00			0.97 (0.54–1.73)	0.91			0.83 (0.46–1.50)	0.54		
Mitotic count $\geq$5 *$^\$$ (*n* = 81)	1.47 (0.62–3.48)	0.38			**2.03 (1.07–3.85)**	**0.03**			**2.36 (1.21–4.63)**	**0.01**		
Necrosis $^\$$	1.25 (0.56–2.81)	0.59			1.15 (0.65–2.06)	0.63			1.15 (0.64–2.07)	0.65		
WHO grade (*n* = 77)		0.29				0.047		0.04		0.03		0.03
Grade 1–2	1				1				1			
Grade 3	1.76 (0.58–5.28)				2.05 (0.98–4.28)		2.14 (1.03–4.78)		2.21 (1.02–4.79)		2.36 (1.08–5.16)	
Max. diameter (*n* = 61)		0.83				0.79				0.74		
< 3 cm	1				1				1			
3–5 cm	1.73(0.18–228.92)				1.96(0.25–15.16)				1.94(0.25–15.00)			
$\geq$5 cm	2.20(0.27–285.32)				1.70(0.22–12.93)				1.55(0.20–11.85)			
Topography *(*n* = 87)		0.004		0.03		0.003				0.002		
Right-sided	1		1		1				1			
Left-sided *	**3.02(1.21–7.52)**		**2.69(1.07–6.81)**		**1.84 (1.01–3.37)**				**2.03 (1.10–3.76)**			
Midline *	**13.86(2.54–75.71)**		**8.29(1.42–48.50)**		**8.57(2.32–31.65)**				**9.08(2.44–33.73)**			
Location		0.94				0.70				0.62		
Supratentorial	1				1				1			
Infratentorial	0.96 (0.32–2.84)				1.15 (0.55–2.40)				1.20 (0.58–2.51)			
Preoperative embolization *$^\$$	0.25 (0.002–1.82)	0.35			**0.08 (0.001–0.55)**	**0.08**			**0.09 (0.001–0.60)**	**0.09**		
Resection grade * (*n* = 84)		**0.04**				**0.12**				**0.10**		**0.03**
Simpson 1–2	1				1		1		1		1	
Simpson 3–4	**3.20 (1.04–9.83)**				**2.02 (0.84–4.89)**				**2.11 (0.87–5.12)**		**3.00 (1.09–8.24)**	
PORT $^\$$ (*n* = 56)	0.88 (0.34–2.29)	0.80			0.50 (0.26–0.97)	0.04			0.56 (0.29–1.10)	0.10		

* Univariate significant parameters (*p* < 0.20) were then tested in a multivariate model; they were significant if *p* < 0.05, indicated by bolded text. HR: hazard ratio; PORT: postoperative radiotherapy; $^\$$ For binary variables, the reference category was "No".

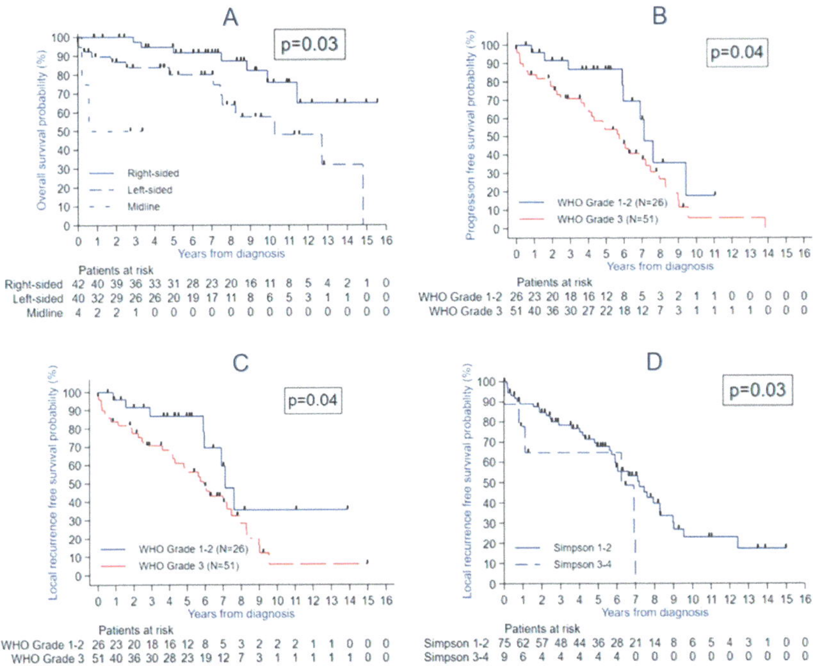

Figure 3. The survival curve obtained with the Kaplan–Meier method showing the difference in OS according to tumour location (**A**), PFS according to the WHO grade (**B**), LRFS according to the WHO grade (**C**), and LRFS according to the extent of resection with newly diagnosed iSFT (**D**) with their respective p value, significant if <0.05.

We observed that treatment influenced the outcome. We observed a significantly lower LRFS in the STR group in the multivariate analyses with HR = 3.00 (95% CI: 1.09–8.24; $p = 0.03$) (Table 3, Figure 3D). However, performing STR was associated with significantly lower OS only in the univariate analysis with HR = 3.20 (95% CI: 1.04–9.83; $p = 0.04$) and tended to be associated with lower PFS (HR = 2.71 (95% CI: 0.99–7.41; $p = 0.052$)). Preoperative embolization was associated with a lower risk of PFS and LRFS in the univariate analysis, but this association was not significant in the multivariate analysis (HR = 0.08 (95% CI: 0.001–0.55; $p = 0.08$) and HR = 0.09 (95% CI: 0.001–0.60; $p = 0.09$), respectively). Adjuvant radiotherapy was significantly associated with greater PFS and LRFS in the univariate analysis (HR = 0.50 (95% CI: 0.26–0.97; $p = 0.04$) and HR = 0.56 [95% CI: 0.29–1.10]; $p = 0.10$)), respectively, but not with OS (HR = 0.88 (95% CI: 0.34–2.29); $p = 0.80$). Sensitivity multivariate analyses were performed with PORT as a covariate. We observed a significantly lower PFS with HR = 0.47 (95% CI: 0.24–0.91; $p = 0.03$), but the association was no more significant regarding LRFS.

4. Discussion

Intracranial SFTs are very rare mesenchymal tumours with a high risk of recurrence [9,15,17]. To the best of our knowledge, this was the largest clinical retrospective cohort describing data at the stage of diagnosis and recurrence in Europe and the USA. There are no established guidelines for the management of iSFTs, a hard-to-treat malignancy, especially at recurrence. Current evidence came from retrospective data and prospective studies with a limited number of cases [18]. We conducted a long-term study permitting the report of 42 patients in relapse and their characterization, which is usually difficult given the rarity of this entity [2,6,19]. Local treatments are often employed early and systemic treatment seems to present a restrained efficiency.

As reported by several authors, we observed that GTR provided a strong benefit in terms of tumour control as a first treatment for iSFTs [6,10,12,13,20]. Kim et al. reported a significant decrease in the 5-year recurrence risk in GTR cases compared to STR (20.8% vs. 72.7%; $p = 0.006$) cases [2]. However, the results of GTR on OS remain controversial. Consistent with our observation, Soyuer et al. found no association between survival and resection extent [10]. However, Rutkowski et al. observed an increased survival with GTR (13 vs. 9.7 years, $p < 0.05$), independent of the realisation of PORT [9].

Several studies have analysed the impact of PORT on iSFT recurrence and survival with conflicting results [7,14,21–23]. In our cohort, PORT seemed to be efficient only in terms of local control and PFS in patients who presented with an aggressive tumour or underwent an incomplete resection as, discussed in Stessin et al. Other authors, such as Coombs et al., even reported improved OS [7,12,14,21]. However, Xiao et al. found no association between PORT and survival or PFS irrespective of the quality of resection [24]. The low percentage of STR cases included in our cohort and missing data induced a lack of power that could explain our observations. Thus, our cohort could not engage the role of adjuvant radiotherapy in these situations, but the results remained promising. Comparison between different radiotherapy techniques and regimens or between stereotactic radiosurgery (SRS) and fractionated radiotherapy was not addressed in our series due to the small number of patients undergoing adjuvant SRS and missing data. Moreover, due to the retrospective nature of our cohort based on surgical specimens, no patient treated with exclusive SRS was included in the study. However, like meningiomas, this treatment strategy should be further explored.

Based on the rationale of the meningioma treatment, a preoperative embolization is sometimes conducted to perform a GTR and limit morbidity in these heavily vascularized tumours [19,25]. To date, its impact on clinical outcome has not been demonstrated in iSFTs. In our series, this technique tended to reduce recurrence. The nonsignificant association in the multivariate models might have been due to the lack of power because of the small number of patients.

Concerning survival predictive factors, age at diagnosis is heavily discussed [10,12,26]. However, as Wang et al., we found that a 1-year increase in age predicted a poor outcome [15]. We observe that the CNS WHO grade III was associated with poorer PFS and LRFS as Macagno et al. had described [4,20].

We reported an association between survival and tumour location. The results concerning midline locations should be interpreted with caution, given the very small number of patients ($n = 4$) with this type of location. To our knowledge, this information was not described in other studies of SFTs. Even if extra- and intra-axial tumours are not similar, some authors have described a poor prognosis in glioblastoma with central and left temporal localizations [27]. The impact of left lateralization on OS in our series might be explained by the greater preoperative risks due to the presence of language centres, frequently located on the left [12,13]. However, we did not observe a negative impact for tumours located at intrasinusal or cerebellar levels, which are usually reported to have worse prognoses [9,12,15].

Very few studies have reported on the treatment and characteristics of patients with iSFTs at diagnosis and at relapse [9,15,16]. To our knowledge, we carried out the first description of the clinical management of relapses in France and the largest one in Europe and in the USA. Recurrences occur mostly at the initial local site and are belated and repetitive [23]. Like Rutkowski et al., we observed that the treatment of the first relapse consisted of local techniques (radiotherapy and surgery) for most of our patients [9]. In concordance with the literature, we observed that the first relapse represented a turning point in iSFT evolution, with reduced recurrence latency over the course of the disease [9,12,15,25].

The specificity of iSFT is its potential for extracranial metastatic spreading in approximately 11% to 60% of cases [1,2,15]. The effectiveness of systemic treatments appears limited [28–30]. Nevertheless, antiangiogenic targeted therapy has shown promising OS and PFS results [8,28,31]. Interestingly, we noticed a long-term recurrence-free survival after the fourth local recurrence in one patient treated using an INF-alpha inhibitor. This observation strengthens the rationale of an influence of angiogenesis inhibition in treating unresectable iSFTs [11,32,33].

Although our cohort consisted of the largest description of iSFT cases in Europe and the USA, several limitations remained, especially those inherent in their retrospective nature. For example, some aspects could not be monitored in our study, especially those related to neuropsychiatric symptoms. Only one iSFT was declared to be revealed with psychiatric relapse, and some others with mental confusion regrouped in high-function disorders, but no psychiatric-specific tests were routinely performed during their surveillance. Nowadays, iSFT diagnosis requires the STAT6 immunohistochemistry status [4,5]. Another limitation of our study was the STAT6 status. The period of inclusion of our study was prior to the current use of this tool, explaining that we had less than 40% of patients with a known STAT6 status in our cohort, but all were positive. Nevertheless, our patients were included if there was a confirmed pathological review of iSFT specimens; otherwise, the patient was excluded ($n = 6$). At relapse, all specimens tested for STAT6 in immunochemistry were positive. Moreover, no difference for patients' characteristics was seen in our cohort depending on a known STAT6 status. We conducted a long-term study that enabled extended surveillance to detect potential late recurrence. However, the time and period of surveillance were not consensual in France [1,9]. This might explain why some of our patients were lost-to-follow-up before 10 years of surveillance. An underestimation of recurrence and metastasis might have occurred.

5. Conclusions

Intracranial SFTs are rare sarcomas, which must be followed closely and for a long time after multimodal treatment because of their slow progression. From this study, there emerged the fact that iSFT management is heterogeneous within and between centres, particularly for the treatment of repeated recurrences.

Potential clinical, biological, and radiological factors predicting recurrence require further in-depth studies, especially at an international level. Moreover, prospective studies are necessary to evaluate new modalities of treatment after GTR or suboptimal resection, such as a new regimen of chemotherapy or refined modern techniques, such as proton therapy.

6. Suggested Recommendations Resulting from Our Clinical Experience

Intracranial SFTs should first be treated with surgical techniques. Adjuvant radiotherapy seemed to reduce recurrence in patients with a high histological grade and incomplete resection. Its indication could be discussed in multidisciplinary meetings. Local recurrence and oligometastatic diseases should be treated with local techniques (surgery, SRS) if feasible. Finally, systemic treatments lack efficiency, and clinical studies should be proposed.

Supplementary Materials: The following supporting information can be downloaded at: https://www.mdpi.com/article/10.3390/cancers15030704/s1, Table S1: initial symptoms in patients with newly diagnosed iSFTs; Table S2: clinical characteristics and medical strategies in patients with distant iSFTs.

Author Contributions: Conceptualization, M.L. (Marine Lottin), A.E., L.B., M.P. and N.P.; Data curation, M.L. (Marine Lottin), A.E., M.A.-T. and M.B. (Maël Barthoulot); Formal analysis, A.E. M.A.-T. and M.B. (Maël Barthoulot); Investigation, M.L. (Marine Lottin), A.E., M.P. and S.Z.; Methodology, M.L. (Marine Lottin), A.E., L.B., M.A.-T., M.B. (Maël Barthoulot), M.P. and N.P.; Project administration, A.E. and N.P.; Resources, L.B., M.B. (Mathieu Boone), S.Z., J.G., G.P., J.P., H.D. E.E., M.L. (Michel Lefranc), S.F., H.N., E.G., J.-J.L., B.M., R.S., S.D., A.M., J.-M.C., C.-A.M. and B.C. Software, L.B., M.A.-T., M.B. (Maël Barthoulot), and S.Z.; Supervision, A.E., B.C. and N.P.; Validation, M.L. (Marine Lottin), A.E., L.B., M.B. (Mathieu Boone), J.G., G.P., J.P., H.D., E.E., M.L. (Michel Lefranc), S.F., H.N., E.G., J.-J.L., B.M., R.S., S.D., A.M., J.-M.C., C.-A.M., B.C. and N.P.; Visualization, N.P. Writing—original draft, M.L. (Marine Lottin) and A.E.; Writing—review and editing, M.L. (Marine Lottin) A.E., L.B., M.A.-T., M.B. (Maël Barthoulot), M.P., M.B. (Mathieu Boone), J.G., G.P., J.P., H.D., E.E. M.L. (Michel Lefranc), S.F., H.N., E.G., J.-J.L., B.M., R.S., S.D., A.M., J.-M.C., C.-A.M., B.C. and N.P All authors have read and agreed to the published version of the manuscript.

Funding: This research received no external funding.

Institutional Review Board Statement: The study was conducted in accordance with the Institutional Review Board of Centre Oscar Lambret. Ethical local approval was obtained on 25 May 2022. The number of the ethical approval is N° 2022-013.

Informed Consent Statement: Informed consent was obtained from all subjects involved in the study

Data Availability Statement: The data presented in this study are available on request from the corresponding author.

Acknowledgments: This study was designed with the participation of the APOTAC "association picarde pour l'optimisation des thérapeutiques anti-cancéreuses" and FBTDB "Recensement national histologique des Tumeurs Primitives du Système Nerveux Central" with the collaboration of the French associations "Association des Neuro-Oncologues d'Expression Française" (ANOCEF) and the "Club de Neuro-Oncologie de la Société Française de Neurochirurgie" (CNO-SFNC). The authors particularly thank the patient associations. The authors thank those who oversaw the FBTDB: Fabienne Bauchet (FBTDB, ICM Montpellier), Valérie Rigau (CHU Montpellier), Amélie Darlix (ICM Montpellier), Brigitte Trétarre (Registre Tumeur Hérault), Montpellier, France; and the nonprofit organizations supporting the FBTDB: Ligue Nationale Contre le Cancer et comité 34, Associations pour la Recherche sur les Tumeurs Cérébrales (ARTC) Nord et Sud, Des étoiles dans la Mer.

Conflicts of Interest: The authors declare no conflict of interest. The funders had no role in the design of the study; in the collection, analyses, or interpretation of data; in the writing of the manuscript, or in the decision to publish the results.

References

1.   Cohen-Inbar, O. Nervous System Hemangiopericytoma. *Can. J. Neurol. Sci. J. Can. des Sci. Neurol.* **2019**, *47*, 18–29. [CrossRef]
2.   Kim, B.S.; Kim, Y.; Kong, D.-S.; Nam, D.-H.; Lee, J.-I.; Suh, Y.-L.; Seol, H.J. Clinical outcomes of intracranial solitary fibrous tumor and hemangiopericytoma: Analysis according to the 2016 WHO classification of central nervous system tumors. *J. Neurosurg.* **2018**, *129*, 1384–1396. [CrossRef]
3.   Darlix, A.; Zouaoui, S.; Rigau, V.; Bessaoud, F.; Figarella-Branger, D.; Mathieu-Daudé, H.; Trétarre, B.; Bauchet, F.; Duffau, H.; Taillandier, L.; et al. Epidemiology for primary brain tumors: A nationwide population-based study. *J. Neuro-Oncology* **2016**, *131*, 525–546. [CrossRef] [PubMed]
4.   Macagno, N.; Vogels, R.; Appay, R.; Colin, C.; Mokhtari, K.; Küsters, B.; Wesseling, P.; Figarella-Branger, D.; Flucke, U.; Bouvier, C.; et al. Grading of meningeal solitary fibrous tumors/hemangiopericytomas: Analysis of the prognostic value of theMarseilleGradingSystem in a cohort of 132 patients. *Brain Pathol.* **2018**, *29*, 18–27. [CrossRef] [PubMed]
5.   Louis, D.N.; Perry, A.; Wesseling, P.; Brat, D.J.; Cree, I.A.; Figarella-Branger, D.; Hawkins, C.; Ng, H.K.; Pfister, S.M.; Reifenberger, G.; et al. The 2021 WHO Classification of Tumors of the Central Nervous System: A summary. *Neuro-Oncology* **2021**, *23*, 1231–1251. [CrossRef] [PubMed]
6.   Kim, J.H.; Jung, H.-W.; Kim, Y.-S.; Kim, C.J.; Hwang, S.-K.; Paek, S.H.; Kim, D.G.; Kwun, B.D. Meningeal hemangiopericytomas: Long-term outcome and biological behavior. *Surg. Neurol.* **2003**, *59*, 47–53, discussion 53. [CrossRef]
7.   Dufour, H.; Métellus, P.; Fuentes, S.; Murracciole, X.; Régis, J.; Figarella-Branger, D.; Grisoli, F. Meningeal Hemangiopericytoma: A Retrospective Study of 21 Patients with Special Review of Postoperative External Radiotherapy. *Neurosurgery* **2001**, *48*, 756–763. [CrossRef]

8.   Martin-Broto, J.; Cruz, J.; Penel, N.; Le Cesne, A.; Hindi, N.; Luna, P.; Moura, D.S.; Bernabeu, D.; de Alava, E.; Lopez-Guerrero, J.A.; et al. Pazopanib for treatment of typical solitary fibrous tumours: A multicentre, single-arm, phase 2 trial. *Lancet Oncol.* **2020**, *21*, 456–466. [CrossRef]

9.   Rutkowski, M.J.; Bloch, O.; Jian, B.J.; Chen, C.; Sughrue, M.E.; Tihan, T.; Barani, I.J.; Berger, M.S.; McDermott, M.W.; Parsa, A.T. Management of recurrent intracranial hemangiopericytoma. *J. Clin. Neurosci.* **2011**, *18*, 1500–1504. [CrossRef]

10.   Soyuer, S.; Chang, E.L.; Selek, U.; McCutcheon, I.E.; Maor, M.H. Intracranial meningeal hemangiopericytoma: The role of radiotherapy: Report of 29 cases and review of the literature. *Cancer* **2004**, *100*, 1491–1497. [CrossRef]

11.   Chamberlain, M.C.; Glantz, M.J. Sequential salvage chemotherapy for recurrent intracranial hemangiopericytoma. *Neurosurgery* **2008**, *63*, 720–727. [CrossRef]

12.   Guthrie, B.L.; Ebersold, M.J.; Scheithauer, B.W.; Shaw, E.G. Meningeal Hemangiopericytoma: Histopathological Features, Treatment, and Long-Term Follow-up of 44 Cases. *Neurosurgery* **1989**, *25*, 514–522. [CrossRef] [PubMed]

13.   Rutkowski, M.J.; Sughrue, M.E.; Kane, A.J.; Aranda, D.; Mills, S.A.; Barani, I.J.; Parsa, A.T. Predictors of mortality following treatment of intracranial hemangiopericytoma: Clinical article. *J. Neurosurg.* **2010**, *113*, 333–339. [CrossRef] [PubMed]

14.   Stessin, A.M.; Sison, C.; Nieto, J.; Raifu, M.; Li, B. The Role of Postoperative Radiation Therapy in the Treatment of Meningeal Hemangiopericytoma—Experience From the SEER Database. *Int. J. Radiat. Oncol. Biol. Phys. Biol. Phys.* **2013**, *85*, 784–790. [CrossRef]

15.   Wang, W.; Zhang, G.-J.; Zhang, L.-W.; Li, D.; Wu, Z.; Zhang, J.-T. Long-Term Outcome and Prognostic Factors After Repeated Surgeries for Intracranial Hemangiopericytomas. *World Neurosurg.* **2017**, *107*, 495–505. [CrossRef] [PubMed]

16.   Schiariti, M.; Goetz, P.; El-Maghraby, H.; Tailor, J.; Kitchen, N.D. Hemangiopericytoma: Long-term outcome revisited. Clinical article. *J. Neurosurg.* **2011**, *114*, 747–755. [CrossRef]

17.   Chen, L.-F.; Yang, Y.; Yu, X.-G.; Gui, Q.-P.; Xu, B.-N.; Zhou, D.-B. Multimodal treatment and management strategies for intracranial hemangiopericytoma. *J. Clin. Neurosci.* **2015**, *22*, 718–725. [CrossRef]

18.   Lottin, M.; Escande, A.; Peyre, M.; Sevestre, H.; Maurage, C.A.; Chauffert, B.; Penel, N. What's new in the management of meningeal solitary fibrous tu-mor/hemoangioperctyoma? *Bull. du Cancer* **2020**, *107*, 1260–1273. [CrossRef]

19.   Jääskeläinen, J.; Servo, A.; Haltia, M.; Wahlström, T.; Valtonen, S. Intracranial hemangiopericytoma: Radiology, surgery, radiotherapy, and outcome in 21 patients. *Surg. Neurol.* **1985**, *23*, 227–236. [CrossRef]

20.   Melone, A.G.; D'Elia, A.; Santoro, F.; Salvati, M.; Delfini, R.; Cantore, G.; Santoro, A. Intracranial Hemangiopericytoma—Our Experience in 30 Years: A Series of 43 Cases and Review of the Literature. *World Neurosurg.* **2013**, *81*, 556–562. [CrossRef]

21.   Combs, S.E.; Thilmann, C.; Debus, J.; Schulz-Ertner, D. Precision radiotherapy for hemangiopericytomas of the central nervous system. *Cancer* **2005**, *104*, 2457–2465. [CrossRef] [PubMed]

22.   Cohen-Inbar, O.; Lee, C.-C.; Mousavi, S.H.; Kano, H.; Mathieu, D.; Meola, A.; Nakaji, P.; Honea, N.; Johnson, M.; Abbassy, M.; et al. Stereotactic radiosurgery for intracranial hemangiopericytomas: A multicenter study. *J. Neurosurg.* **2017**, *126*, 744–754. [CrossRef] [PubMed]

23.   Lee, J.H.; Jeon, S.H.; Park, C.K.; Park, S.H.; Yoon, H.I.; Chang, J.H.; Suh, C.O.; Kang, S.J.; Lim, D.H.; Kim, I.A.; et al. The Role of Postoperative Radiotherapy in Intracranial Solitary Fibrous Tumor/Hemangiopericytoma: A Multi-Institutional Ret-rospective Study (KROG 18-11). *Cancer Res. Treat.* **2022**, *54*, 65–74. [CrossRef] [PubMed]

24.   Xiao, J.; Xu, L.; Ding, Y.; Wang, W.; Chen, F.; Zhou, Y.; Zhang, F.; Zhou, Q.; Wu, X.; Li, J.; et al. Does post-operative radiotherapy improve the treatment outcomes of intracranial hemangiopericytoma? A retrospective study. *BMC Cancer* **2021**, *21*, 1–10. [CrossRef] [PubMed]

25.   Fountas, K.N.; Kapsalaki, E.; Kassam, M.; Feltes, C.H.; Dimopoulos, V.G.; Robinson, J.S.; Smith, J.R. Management of intracranial meningeal hemangiopericytomas: Outcome and experience. *Neurosurg. Rev.* **2006**, *29*, 145–153. [CrossRef] [PubMed]

26.   Kumar, N.; Kumar, R.; Kapoor, R.; Ghoshal, S.; Kumar, P.; Salunke, P.S.; Radotra, B.D.; Sharma, S.C. Intracranial meningeal hemangiopericytoma: 10 years experience of a tertiary care Institute. *Acta Neurochir.* **2012**, *154*, 1647–1651. [CrossRef]

27.   Fyllingen, E.H.; Bø, L.E.; Reinertsen, I.; Jakola, A.S.; Sagberg, L.M.; Berntsen, E.M.; Salvesen, Ø.; Solheim, O. Survival of glioblastoma in relation to tumor location: A statistical tumor atlas of a population-based cohort. *Acta Neurochir.* **2021**, *163*, 1895–1905. [CrossRef]

28.   Park, M.S.; Patel, S.R.; Ludwig, J.A.; Trent, J.C.; Conrad, C.A.; Lazar, A.J.; Wang, W.-L.; Boonsirikamchai, P.; Choi, H.; Wang, X.; et al. Activity of temozolomide and bevacizumab in the treatment of locally advanced, recurrent, and metastatic hemangiopericytoma and malignant solitary fibrous tumor. *Cancer* **2011**, *117*, 4939–4947. [CrossRef]

29.   Bastin, K.T.; Mehta, M.P. Meningeal hemangiopericytoma: Defining the role for radiation therapy. *J. Neuro-Oncology* **1992**, *14*, 277–287. [CrossRef]

30.   Beadle, C.F.; Hillcoat, B.L. Treatment of advanced malignant hemangiopericytoma with combination adriamycin and dtic: A report of four cases. *J. Surg. Oncol.* **1983**, *22*, 167–170. [CrossRef]

31.   De Lemos, M.L.; Kang, I.; Schaff, K. Efficacy of bevacizumab and temozolomide therapy in locally advanced, recurrent, and metastatic malignant solitary fibrous tumour: A population-based analysis. *J. Oncol. Pharm. Pract.* **2018**, *25*, 1301–1304. [CrossRef] [PubMed]

32. Lackner, H.; Urban, C.; Schwinger, W.; Kerbl, R.; Sovinz, P. Interferon alfa-2a in recurrent metastatic hemangiopericytoma. *Med Pediatr. Oncol.* **2003**, *40*, 192–194. [CrossRef] [PubMed]
33. Kirn, D.H.; Kramer, A. Long-Term Freedom From Disease Progression With Interferon Alfa Therapy in Two Patients With Malignant Hemangiopericytoma. *Gynecol. Oncol.* **1996**, *88*, 764–765. [CrossRef] [PubMed]

Article

Trabectedin Is Active against Two Novel, Patient-Derived Solitary Fibrous Pleural Tumor Cell Lines and Synergizes with Ponatinib

Bahil Ghanim [1,2,3], Dina Baier [2,4,5,*], Christine Pirker [2], Leonhard Müllauer [6], Katharina Sinn [1], Gyoergy Lang [1], Konrad Hoetzenecker [1] and Walter Berger [2,5]

1 Department of Thoracic Surgery, Medical University of Vienna, 1090 Vienna, Austria
2 Center for Cancer Research and Comprehensive Cancer Center, Medical University of Vienna, 1090 Vienna, Austria
3 Department of General and Thoracic Surgery, Karl Landsteiner University of Health Sciences, University Hospital Krems, 3500 Krems, Austria
4 Institute of Inorganic Chemistry, University of Vienna, 1090 Vienna, Austria
5 Research Cluster "Translational Cancer Therapy Research", 1090 Vienna, Austria
6 Department of Pathology, Medical University of Vienna, 1090 Vienna, Austria
* Correspondence: dina.baier@univie.ac.at

Simple Summary: Solitary fibrous tumor of the pleura (SFT) is an orphan disease resistant to standard systemic therapy. We managed to establish two patient-derived cell models characterized as SFT by the *NAB2-STAT6* gene fusion. Cell lines were tested for drug responsiveness in vitro. Trabectedin and distinct multi-tyrosine kinase inhibitors were effective as single agents. Most interestingly, the combination of trabectedin with ponatinib or dasatinib showed synergistic effects against fusion-positive SFT cell viability, thus suggesting two novel, potentially interesting treatment regimens for this rare and, to date, treatment-refractory disease.

Citation: Ghanim, B.; Baier, D.; Pirker, C.; Müllauer, L.; Sinn, K.; Lang, G.; Hoetzenecker, K.; Berger, W. Trabectedin Is Active against Two Novel, Patient-Derived Solitary Fibrous Pleural Tumor Cell Lines and Synergizes with Ponatinib. *Cancers* **2022**, *14*, 5602. https://doi.org/10.3390/cancers14225602

Academic Editor: Nicola Baldini

Received: 10 September 2022
Accepted: 10 November 2022
Published: 15 November 2022

Publisher's Note: MDPI stays neutral with regard to jurisdictional claims in published maps and institutional affiliations.

Abstract: Solitary fibrous tumor of the pleura (SFT) is a rare disease. Besides surgery combined with radiotherapy in nondisseminated stages, curative options are currently absent. Out of fourteen primo-cell cultures, established from surgical SFT specimens, two showed stable in vitro growth. Both cell models harbored the characteristic *NAB2-STAT6* fusion and were further investigated by different preclinical methods assessing cell viability, clone formation, and protein regulation upon single-drug treatment or in response to selected treatment combinations. Both fusion-positive cell models showed—in line with the clinical experience and the literature—a low to moderate response to most of the tested cytotoxic and targeted agents. However, the multi-tyrosine kinase inhibitors ponatinib and dasatinib, as well as the anti-sarcoma compound trabectedin, revealed promising activity against SFT growth. Furthermore, both cell models spontaneously presented strong FGFR downstream signaling targetable by ponatinib. Most interestingly, the combination of either ponatinib or dasatinib with trabectedin showed synergistic effects. In conclusion, this study identified novel trabectedin-based treatment combinations with clinically approved tyrosine kinase inhibitors, using two newly established *NAB2-STAT6* fusion-positive cell models. These findings can be the basis for anti-SFT drug repurposing approaches in this rare and therapy-refractory disease.

Keywords: solitary fibrous tumor; *NAB2-STAT6* gene fusion; targeted therapy; trabectedin; ponatinib; dasatinib; in vitro; patient-derived cell lines

1. Introduction

Solitary fibrous tumor of the pleura (SFT) belongs to the group of soft tissue tumors and can be regarded as an orphan disease with an annual incidence of less than 0.1 per 100,000 in Europe [1]. SFT is defined as mesenchymal neoplasm and was previously divided into a benign and malignant subtype by the England criteria [2]. In 2002, de Perrot et al. suggested a staging system including macroscopic and histologic growth patterns [3]. Both the tumor dignity and the de Perrot staging proved to estimate the probability of recurrence [4]. Nevertheless, the most recent, fifth edition of the WHO classification suggests to avoid the terms benign and malignant and instead use risk classification models since the previous histological stratification did not accurately reflect the clinical behavior [5]. Regarding the genetic characterization, the *NAB2-STAT6* gene fusion was described as a distinct hallmark of SFT. The fusion protein promotes the malignant phenotype via constitutive activation of early growth response 1 (EGR1)-mediated gene transcription by inhibiting the EGR1-repressing activity of wild-type NAB2 [6,7]. In addition to the fusion transcript detection, nuclear STAT6 staining can be useful in the diagnosis of SFT besides the traditional immunohistochemical markers vimentin, CD34, CD99, and bcl-2 [4,8]. Moreover, fibroblast growth factor receptor 1 (FGFR1) and its ligand fibroblast growth factor 2 (FGF2) are overexpressed in SFT [9,10], the ligand being even suggested as a prognostic parameter [11].

The clinical outcome after resection with a curative intention for SFT is—compared to other thoracic malignancies—excellent, with 77% and 67% of patients being still alive without evidence of disease after 5 and even 10 years, respectively, as demonstrated before by our study group in a large international multicenter study analyzing 125 pleural SFT patients [12]. Today, radical surgery is the only treatment modality able to cure the disease and chemo- or radiotherapy alone failed to distinctly improve clinical outcome, resulting in a significant lack of noninvasive treatment alternatives for patients presenting with unresectable SFT, i.e., due to enhanced tumor size or dissemination [13]. In addition, even completely resected tumors can recur decades after radical resection and not all patients are eligible candidates for thoracic surgery or redo surgery at the time of later recurrence. Thus, improving the systemic treatment of SFT, either as part of multimodality therapy concepts including surgery or used alone to treat the unresectable disease, is urgently warranted [13–15].

With regard to systemic therapy, most studies were—according to the low incidence of SFT—of a retrospective nature and analyzed only small numbers of patients [16]. In these studies, low to moderate response rates were reported for conventional chemo-, as well as targeted therapy [14,15,17–19]. Based on this knowledge, we aimed to establish patient-derived SFT cell models to test modern anticancer compounds and their combinations in vitro in frame of a drug repurposing approach for this otherwise treatment-resistant orphan disease. To acknowledge the most recent major breakthrough with regard to the *NAB2-STAT6* fusion characterizing SFT as a translocation-related soft tissue tumor, we focused on therapy approaches that proved to previously be efficient in translocation- and fusion-related malignant diseases [5].

2. Materials and Methods

2.1. Establishment of Patient-Derived Cell Lines

All patients with a clinical SFT diagnosis receiving tumor resection between 12/2014 and 12/2021 at the Medical University of Vienna, Department of Thoracic Surgery were included. Tumor samples were retrieved during surgery and used for cell culturing after written consent was granted by each included patient. The study was approved by the Ethics Committee of the Medical University of Vienna under the title, "Etablierung einer Tumorbank für die molekulare Analyse von thorakalen Tumoren", EK NR.: 9004/2009. In brief, a small part of the tumor was removed in the operating room for tumor cell isolation and brought to the Center for Cancer Research and Comprehensive Cancer Center, Medical University of Vienna. The tumor probe was shredded and then grown in ACL-4 medium (ATCC MEDIUM 8002) supplemented with human epidermal growth factor (1 ng/mL, E9644), hydrocortisone (10 µg/mL, T2036), transferrin (24.2 ng/mL, H2270), and fetal calf serum (10%). Penicillin (100 U/mL) and streptomycin (10 mg/mL), all obtained from Sigma-Aldrich (St. Louis, MO, USA), were added to the medium for primary culture establishment. The cells were cultured in a humidified incubator at 37 °C and 5% CO_2.

After demonstrating stable in vitro growth, the cell lines were further investigated to prove their SFT origin, as later described in the Results section.

2.2. Immunohistochemistry

Routine histopathological processing, including immunohistochemistry (IHC), of the patients' tumor tissues was performed at the Department of Pathology of the Medical University of Vienna. The antibody panels used for immunohistochemistry included bcl-2 (clone 124; DAKO Agilent, Santa Clara, CA, USA), CD34 (clone QBend/10; Leica, Wetzlar, Germany), CD99 (clone EP8, Biocare, Pacheco, CA, USA), STAT6 (anti-Stat 6 polyclonal antibody; Sigma-Aldrich, St. Louis, MO, USA), and Ki67 (clone 30-9; Ventana, Oro Valley, AZ, USA). In addition, all cases were re-reviewed by the author LM to confirm pathologic SFT diagnosis.

2.3. Detection of the NAB2-STAT6 Fusion

RNA was isolated from formalin-fixed and paraffin-embedded tumor samples (FFPE) and the corresponding cell lines. The *NAB2-STAT6* fusion was analyzed by using next-generation sequencing (NGS). For both materials (FFPE tissue and corresponding cell lines), the TruSight RNA Fusion Panel (Illumina, San Diego, CA, USA) was used for library generation. Sequencing was performed with a MiSeq instrument (Illumina, San Diego, CA, USA). All NGS results were verified by reverse transcription PCR as described below in Section 2.9.

2.4. Drugs

Ponatinib was purchased from LC Laboratories (Woburn, MA, USA). Trabectedin was obtained from PharmaMar (Colmenar Viejo, Spain). Cisplatin was kindly provided by the Institute of Inorganic Chemistry, University of Vienna. Ponatinib, dasatinib, obatoclax mesylate, venetoclax, vincristine, PD173074, and stattic were obtained from Selleckchem (EUBIO, ANDREAS KÖCK e.U., Vienna, Austria). Nintedanib and imatinib were purchased from LC Laboratories (Woburn, MA, USA). Doxorubicin was acquired from Sigma-Aldrich (St. Louis, MO, USA). Paclitaxel was procured from Bristol Myers Squibb (New York City, NY, USA).

2.5. Determination of Cell Proliferation

For the determination of cell proliferation, SFT-T1, SFT-T2, HCC827, and Met5a cells were seeded at a density of 3.5×10^3 cells per well in triplicates in 96-well plates in 300 μL of ACL medium or RPMI-1640 (Sigma-Aldrich, St. Louis, MO, USA) supplemented with 10% fetal calf serum and incubated overnight. Cells were imaged the next day (T0), and 24, 48, and 72 h later using the Cytation 5 Cell Imaging Multimode Reader (BioTek, as part of Agilent, Winooski, VT, USA). A digital phase contrast was created from the derived images. Following this, pixel intensities per well were quantified by using ImageJ 1.50i (NIH, Bethesda, MD, USA).

2.6. Cell Viability Assay

The 3-(4,5-dimethylthiazol-2-yl)-2,5-diphenyltetrazolium bromide (MTT) assay (EZ4U, Biomedica, Vienna, Austria) was utilized as a cell viability and drug sensitivity assay to screen for the in vitro responsiveness of our cell lines. SFT cell lines were seeded at a density of 3.5×10^3 cells/100 μL in 96-well microtiter plates and incubated overnight. Cells were exposed to the respective single drugs and drug combinations for 72 h. Cell viability was measured as published before [20]. From full dose–response curves, the respective IC_{50} values were calculated by nonlinear regression curve-fitting (sigmoidal dose–response with variable slope). To evaluate the efficacy of drug combinations, the combination index (CI) was calculated according to the method published by Chou and Talalay [21] using the CalcuSyn software (Biosoft, Ferguson, MO, USA).

2.7. Clonogenicity (Clone Formation) Assay

For clonogenic assays, cells were seeded at a density of 1000 cells per well in 500 μL of medium in 24-well microtiter plates and allowed to adhere overnight. Cells were exposed to the indicated single drugs and combinations for nine days. After washing with phosphate-buffered saline (PBS, pH 7.4; Sigma-Aldrich Inc., St. Louis, MO, USA), cells were fixed with acetone, and stained with 0.01% (*w/v*) crystal violet. Crystal violet was re-solubilized with 2% sodiumdodecylsulfate (Sigma-Aldrich, St. Louis, MO, USA) and the fluorescence intensity was measured on a spectrophotometer (Tecan Infinite 200 Pro, Tecan Trading AG, Männedorf, Switzerland).

2.8. Western Blot Analyses

For Western blot analyses, cells were seeded into 6-well plates at a density of 4×10^5 cells per well in 2 mL of medium and incubated overnight. Cells were exposed to the respective drugs and combinations for the indicated time, and thereafter, proteins were harvested and analyzed by gel electrophoresis and immunoblotting as published before [20]. FGF2 was acquired from PeproTech (FGF2, Cranbury, NJ, USA). For FGF2 stimulation, cells were starved (medium without FBS) for 24 h prior to treatment. Cells were treated with the indicated drug concentrations for 1 h. Then, 15 min before the end of the treatment time, FGF2 was added for 15 min at a concentration of 20 ng/mL. Sample collection, protein isolation, separation, and Western blotting were performed as described previously [22]. Primary antibodies phospho-p44/42 MAPK (p-Erk) (Thr202/Tyr204) (20G11) (#4376, dilution 1:1000), Erk1/2 (#4695, 1:1000), p-Akt (Ser473) (D9E) (#4060, 1:500), Akt (pan) (C67E7) (#4691, 1:1000), p-Src (Tyr416) (#6943, 1:500), Src (36D10) (#2109, 1:1000), p-S6 ribosomal protein (Ser240/244) (D68F8) (#5364, 1:1000), STAT6 (D3H4) (#5397, 1:1000), and vimentin (R28) (#3932, 1:1000) were purchased from Cell Signaling Technology (Danvers, MA, USA). Anti-Flg (FGFR1; C-15) (sc-121, 1:250) and S6 (C-8) (sc-74459, 1:1000) were acquired from Santa Cruz Biotechnology (Dallas, TX, USA). Anti-ß-actin (AC-15) (A5441, 1:2000) was obtained from Sigma-Aldrich (St. Louis, MO, USA). Horseradish peroxidase-linked secondary antibodies, anti-mouse IgG (Fc specific) antibody (A0168) and anti-rabbit IgG antibody (7074S) were purchased from Merck KGaA (Darmstadt, Germany) and Cell Signaling Technology (Danvers, MA, USA).

2.9. RNA Isolation, Reverse Transcription into cDNA, and RT-PCR

Total RNA was isolated with Trizol using standard protocols and reverse transcribed into cDNA as described in Berger et al. [23]. An RT-PCR was performed to screen for the presence of the *NAB2-STAT6* fusion gene using the primer sequences (Set 1) and the PCR conditions described by Guseva et al. [24]. *GAPDH* served as a housekeeping gene (*GAPDH fw*: 5′-CTG GCG TCT TCA CCA CCA T-3′; *GAPDH rev*: 5′-GCC TGC TTC ACC ACC TTC T-3′).

2.10. Statistical Analyses

Metric data are given as mean $\pm$ SD if not indicated otherwise. An unpaired Student's *t*-test was used to detect significant differences between two groups. A one-way analysis of variance (ANOVA) was used to compare means of more than two groups, and the Bonferroni or Tukey post-test were utilized to correct for multiple testing. *p*-values below 0.05 were considered statistically significant: * $p < 0.05$, ** $p < 0.005$, *** $p < 0.0005$, and **** $p < 0.0001$. All calculations and graphs were performed with Prism 8.0 software (GraphPad, San Diego, CA, USA).

3. Results

3.1. Establishment of Novel SFT Models from Surgical Specimens

In brief, fourteen primo-cell cultures were derived from patients with clinical SFT suspicion during their surgical tumor resection between 2014 and 2021. To prove that all included 14 cell lines were of SFT origin, we first reviewed postoperative pathology reports of the respective patients. Four cell lines had to be excluded because the final histology did not confirm clinical SFT suspicion. Out of the remaining ten cell lines, five turned out to be senescent or did not grow stably in vitro. For the final five stable-growing cell lines, the corresponding patient tumor tissue of the pathology archive was re-evaluated and tested for the *NAB2-STAT6* fusion by NGS to confirm the clinical and histological SFT diagnosis on a genetic level. In three of these cell lines, the genomic fusion event was missing, despite its presence in the original surgical specimen. Thus, finally, two patient-derived SFT cell models stably expressing the *NAB2-STAT6* fusion protein were established (Figure 1), reflecting, on the one hand, the rarity of the disease, and on the other hand, the challenge to culture SFT in vitro. One cell line was derived from the low-risk group SFT (SFT-T1) and the other from the intermediate-risk group SFT (SFT-T2), according to the Demicco classification [25], as summarized in Table 1.

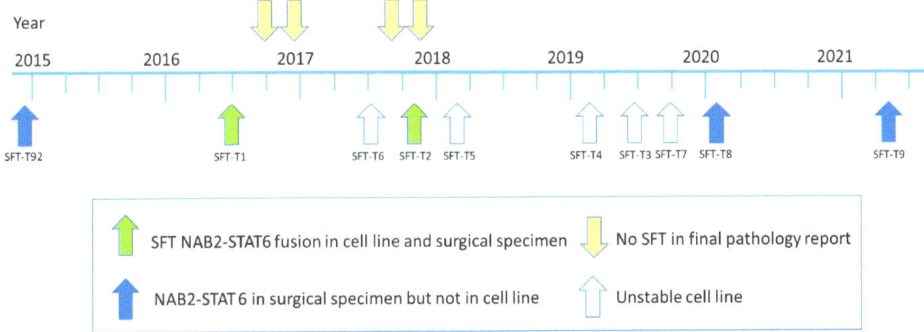

Figure 1. Timeline of study in- and exclusion. Out of 14 cell lines included during the seven-year project period, 2 were finally suitable to be used for the following experiments. In both, the patients' tumor material as well as the cell lines, the NAB2-STAT6 fusion was detected and, accordingly, SFT origin was proven on the genetic and histologic level.

Table 1. Clinical, pathological, and genetic characterization of the study population and the respective cell lines.

SFT#	Age	Sex	Risk Group	CD34	bcl-2	CD99	STAT6	Ki67	*NAB2-STAT6*	*NAB2-STAT6*
							Tumor			**Cell Line**
T92	72	male	intermediate	+	+	+	+	15	+	−
T1	50	male	low	+	+	+	+	1	+	+
T2	45	female	intermediate	+	+	+	+	5	+	+
T8	34	female	intermediate	+	+	+	+	20	+	−
T9	76	female	high	+	+	+	+	25	+	−

Abbreviations: +—positive; −—negative; Ki67—% of positive tumor cells; risk group according to Demicco et al. [25].

Representative pictures of the surgical tumor specimens from SFT-T1 and SFT-T2 showing the general microscopic growth pattern and the IHC profile are given in Figure 2. Both tumors were (immune-)histologically fitting well to the pathological diagnosis of SFT (Table 1). Tumor cells were positive with regard to the membranous and cytoplasmic expression of CD34, bcl-2, and CD99 and furthermore showed nuclear STAT6 positivity (Figure 2b,c,d, and f, respectively). Fibromatosis (Fibr) served as a pathological negative control, and STAT6 and CD34 staining was absent in the respective tissue sample. Furthermore, negative control staining with isotypes of the corresponding antibodies can be found in Figure S1. Expression of proliferation marker Ki67 (Figure 2e) is lower in the low-risk group representative SFT-T1 and fibromatosis tissue as compared to that in intermediate-risk SFT-T2. The corresponding quantification of Ki67 is given in Table 1.

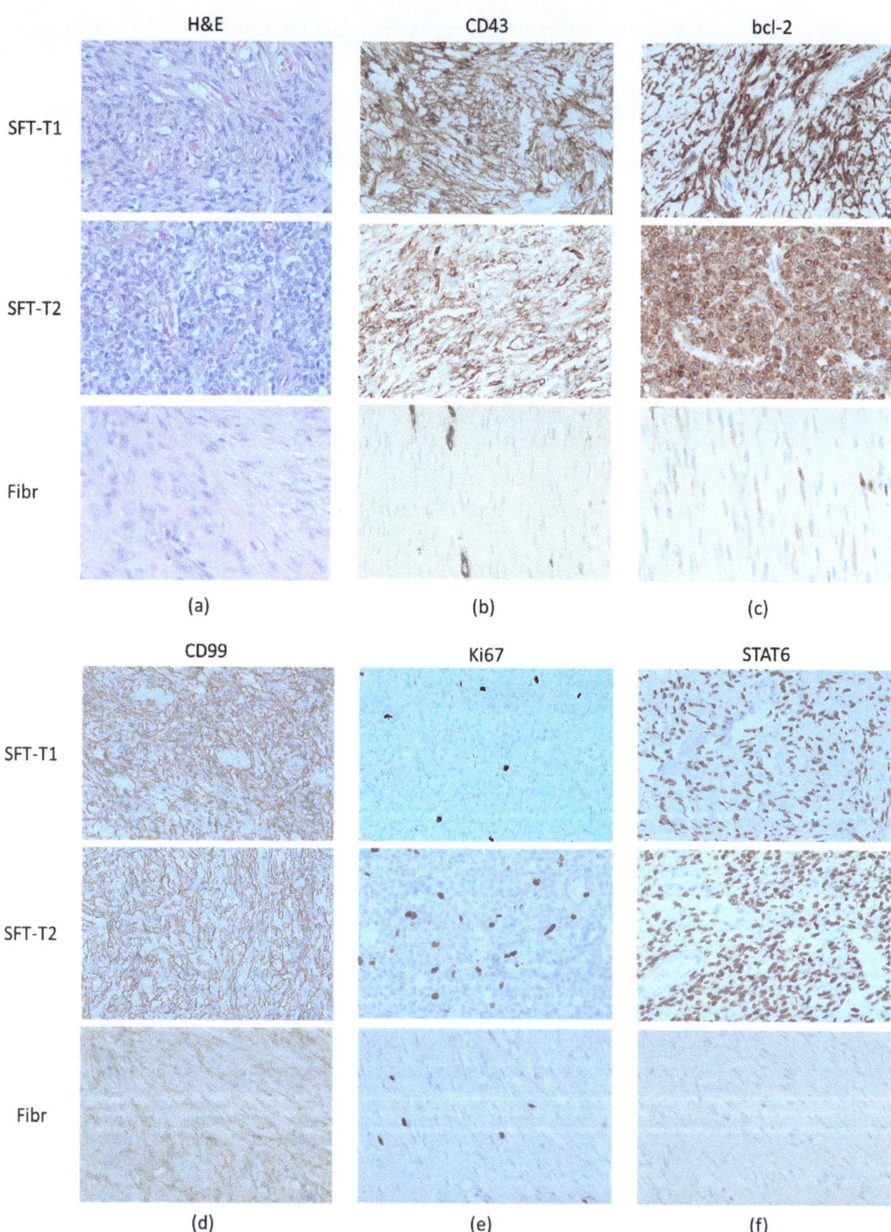

Figure 2. Representative (immune-)histological pictures of the two included tumors (SFT-T1 and SFT-T2) and an excluded fibromatosis patient (Fibr). The patient tumor samples showed the SFT typical pathomorphologic architecture in the H&E staining (**a**) and also the immunohistochemistry was positive for the diagnostic markers CD34 (**b**), bcl-2 (**c**), CD99 (**d**), and nuclear STAT6 staining (**f**), proving that both cell lines (SFT-T1 and -T2) were derived from SFT patients on the pathologic level. In addition, the low-risk group SFT-T1 showed less Ki67 (**e**) expression compared to its intermediate-risk group counterpart. In contrast to the SFT tumor tissue, fibromatosis was negative with regard to CD34 and STAT6 expression.

Both cell lines demonstrated stable in vitro proliferation (Figure S2) and a mesenchymal growth pattern compared to non-small-cell lung cancer HCC827 and pleural mesothelial Met5a cells with epithelial morphology, as shown in Figure 3a. Furthermore, SFT marker expression was validated by Western blot analyses in vitro. Expression of the SFT marker proteins bcl-2 and vimentin (Figure 3b, original Western blots Figure S6) in the cell models reflected the immunohistochemical results of the original patient material (compare Figure 2) and the pathological SFT diagnosis as shown in Table 1. In addition, the presence of the fusion oncogene detected by NGS was verified on the RNA level by reverse transcription PCR (Figure 3c). Fitting to the predicted size of the *NAB2-STAT6* fusion protein, an aberrant Western blot STAT6 signal was detected at 135 kDa in SFT-T1 and SFT-T2 as compared to the wild-type STAT6 signal at 110 kDa expressed by the immortalized pleura cell line Met5a and non-small cell lung cancer HCC827 cells (Figure 3d, original Western blots Figure S6).

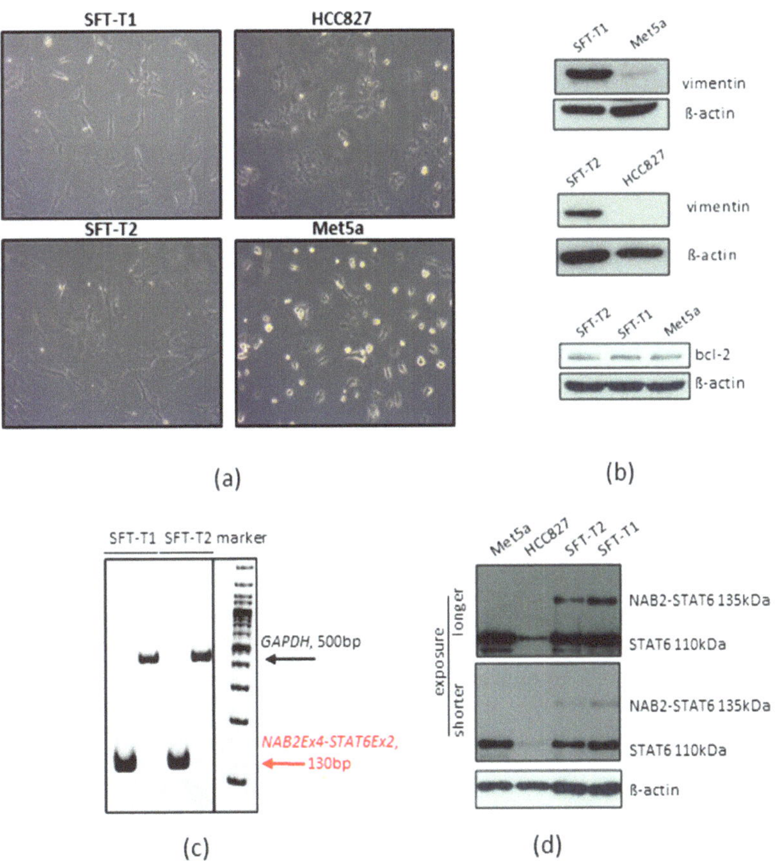

Figure 3. Growth and SFT marker protein expression pattern of SFT-T1 and SFT-T2 cells. (**a**) SFT typical mesenchymal tumor cell shape as compared to HCC827 and Met5a. (**b**) Protein expression concerning vimentin and bcl-2 in SFT-T1 and SFT-T2 cells as compared to Met5a and HCC827 cells used as negative and positive controls, respectively. ß-actin served as loading control. (**c**) Verification of NAB2-STAT6 expression by RT-PCR. GAPDH served as loading control. (**d**) Expression of wild-type and NAB2-STAT6 fusion protein in SFT and control cells as indicated.

Interestingly, NGS analysis of both cell lines and the tumor tissue from all five patients exhibited the identical *NAB2-STAT6* fusion (NAB2 chr12:57486749:+ STAT6 chr12:57502081:-;

NAB2 exon 4/STAT6 exon 2; HG19), as demonstrated in Figure 4. As previously published, this variant of the *NAB2-STAT6* fusion is the most prevalent alteration in pleural SFT [26].

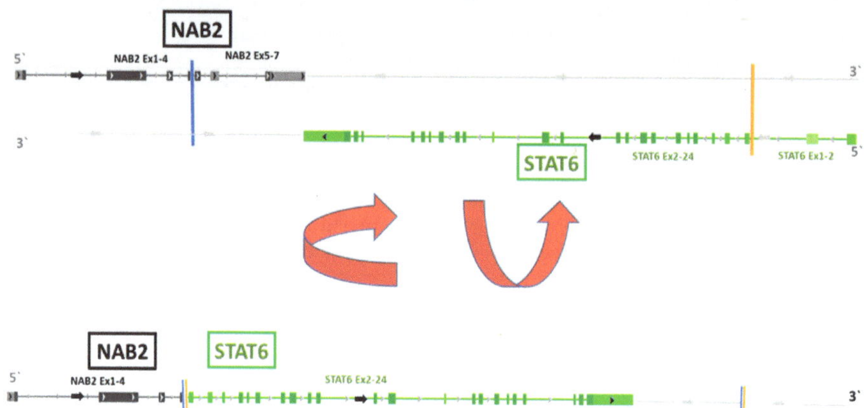

Figure 4. NAB2-STAT6 fusion variant as detected in all four SFT cell models by NGS. The lower part of the figure depicts the NAB2-STAT6 fusion variant (NAB2 exon 4, Stat6 exon 2) obtained by inversion of individual NAB2 and STAT6 strands shown in the upper part. The fusion gene was present in both cell lines (SFT-T1 and SFT-T2) and in all five surgically removed tumor samples, as found by NGS and PCR investigations. This fusion variant is associated with pleural SFT as published [26].

3.2. Screening for Treatment Responses in Novel SFT Cell Models

After establishing the two SFT cell models with a confirmed *NAB2-STAT6* fusion, we first screened for the efficacy of different chemotherapeutic and targeted compounds to obtain a general idea of SFT-specific in vitro drug responsiveness. Representative drug–response curves are provided in Figure S3. The response to the widely used anthracycline doxorubicin was limited in both, the cell line derived from the low-risk SFT (SFT-T1) and the one derived from the intermediate-risk SFT (SFT-T2). This is in line with the literature, where the response to doxorubicin was limited to the dedifferentiated subtype, as reviewed before [16]. Similar results were found for other commonly used cytotoxic compounds including cisplatin and taxol, as summarized in Table 2.

Table 2. In vitro drug responsiveness of two SFT cell models including respective main targets of the drugs.

	IC$_{50}$ ± SD		
Drug	**SFT-T1**	**SFT-T2**	**Main Target(s)**
Cisplatin (μM)	16.38 ± 1.91	14.64 ± 4.91	DNA synthesis
Paclitaxel (μM)	0.032 ± 0.04	0.451 ± 0.57	cytoskeletal components
Vincristine (nM)	37.50 ± 9.09	36.54 ± 10.08	microtubule formation
Venetoclax (μM)	27.91 ± 5.34	18.03 ± 4.64	bcl-2
Obatoclax (nM)	258.55 ± 84.24	237.69 ± 33.68	bcl-2
Doxorubicin (nM)	555.36 ± 327.33	410.51 ± 193.32	DNA topoisomerase II
Imatinib (μM)	27.28 ± 6.92	18.18 ± 2.70	v-Abl, c-Kit, and PDGFR
Stattic (μM)	3.04 ± 1.09	5.30 ± 1.52	STAT3
Trabectedin (nM)	4.36 ± 0.73	3.54 ± 0.98	DNA damaging, blocks DNA binding of FUS-CHOP
Ponatinib (μM)	0.47 ± 0.07	1.89 ± 0.98	FGFR, Abl, PDGFRα, VEGFR2, and Src
Nintedanib (μM)	3.59 ± 0.89	3.63 ± 1.84	VEGFR1-3, FGFR1-3, PDGFR α, and β
Dasatinib (μM)	0.39 ± 0.02	0.40 ± 0.19	Abl, Src, and c-Kit
PD173074 (μM)	2.32 ± 0.73	2.46 ± 0.46	FGFR1 and VEGFR2

Abbreviations: IC$_{50}$—half maximal inhibitory concentration; SD—standard deviation.

From the tested drugs reported thus far, DNA-interfering trabectedin and the multi-tyrosine kinase inhibitors (TKIs) ponatinib, dasatinib, and nintedanib, as well as the more FGFR1-specific PD173074, showed pronounced activity against SFT tumor growth (Table 2). Furthermore, the bcl-2 inhibitor obatoclax was moderately active when compared to, e.g., colorectal cancer cell lines [27] or bladder cancer [28]. Of note, the highly efficient compounds ponatinib and dasatinib, as well as trabectedin, are well known to be active against translocation-/fusion-related malignant diseases as summarized before [29,30], and thus might also be suitable therapies against fusion-positive SFT.

3.3. Ponatinib and Trabectedin Are Active against SFT Cell Growth and Synergize In Vitro

Trabectedin demonstrated high anti-SFT activity in both tested cell models as a single-agent treatment with an IC_{50} in the low nM range. Ponatinib was also highly effective as a single drug with low IC_{50} values in the high nM to low µM range. Respective dose–response curves are given in Figure 5a,b. Upon combination of ponatinib and trabectedin, a strong synergistic effect of this regimen was observed in cell viability assays with CI values < 0.9 (Figure 5c,d).

The synergistic effect of trabectedin and ponatinib was, furthermore, tested in clone formation assays for a longer treatment duration (Figure 5e,f). In SFT-T2 cells, clone formation was significantly reduced by ponatinib and trabectedin as single compounds. Upon combination of the two drugs, the clone formation capacity was further inhibited at low doses of both drugs, verifying the observations from short-term cell viability testing. In SFT-T1 cells, comparable results were obtained (Figure S4).

3.4. Ponatinib Treatment Targets Fibroblast Growth Factor Receptor Downstream Signaling in SFT Cells

For an in-depth evaluation of the mode of action of ponatinib in SFT-T1 and SFT-T2 cells, Western blot analyses were performed (Figure 6, original Western blots Figure S6). The expression of the FGFR downstream signaling proteins of the MAPK and PI3K/Akt pathway [31], as well as their activating phosphorylation, was analyzed after exposure to ponatinib [30]. Since the *NAB2-STAT6* fusion is linked to the activation of the FGFR signaling cascade [5], we also analyzed the effects of FGF2 stimulation on its downstream targets in SFT with and without FGFR inhibition. Of note, SFT responded strongly to FGF2 exposure by activating both the PI3K/Akt and MAPK pathway, indicated by Akt and Erk phosphorylation, respectively. This suggests an important role of FGF signaling in SFT as hypothesized [5] and demonstrated in SFT patients before [11]. Importantly, ponatinib efficiently inhibited both basal and FGF2-induced FGFR downstream signaling pathways, represented by the loss of phosphorylation of Erk and Akt. The complete blockade of the PI3K/Akt pathway was confirmed at the level of the downstream mediator S6. Of note, SFT-T1 cells exhibited phosphorylation of S6 without FGF2 stimulation in contrast to SFT-T2, which showed no basal S6 phosphorylation and less basal Akt phosphorylation. Additionally, basal FGFR1 expression was stronger in SFT-T2 compared to SFT-T1 cells. Surprisingly, the phosphorylation of Src (also, besides FGF signaling, a ponatinib target as reviewed before in [30]) was unaffected by the FGFR inhibitor in our fusion-positive SFT cells. In addition, Src expression was slightly enhanced after ponatinib exposure, indicating a potential dose- or specificity-dependent impact on the Src kinase as suggested before [32].

3.5. Dasatinib Is Active against SFT Growth In Vitro and Enhances Trabectedin Activity

Similar to ponatinib, dasatinib showed significant preclinical activity against our two fusion-positive cell models in the nM range (compare Table 2). A representative dose–response curve is depicted in Figure 7a. However, the first clinical data about dasatinib activity against sarcoma growth that included 25 SFT patients were disappointing [33]. Nevertheless, our observations suggest that dasatinib is active against SFT in vitro. Furthermore, the combination of dasatinib with trabectedin proved additive to synergistic efficacy in cell viability assays as demonstrated by CI-values below 1.2 or 0.9, respectively

(Figure 7b,c)). In line with these findings, the dasatinib/trabectedin combination also further decreased clone formation as compared to the single treatments alone over a longer treatment duration (Figure 7d), suggesting that the response of SFT toward dasatinib can be enhanced by trabectedin. This finding in the SFT-T2 models corresponded well with observations in the SFT-T1 cells at a higher trabectedin concentration (Figure S5).

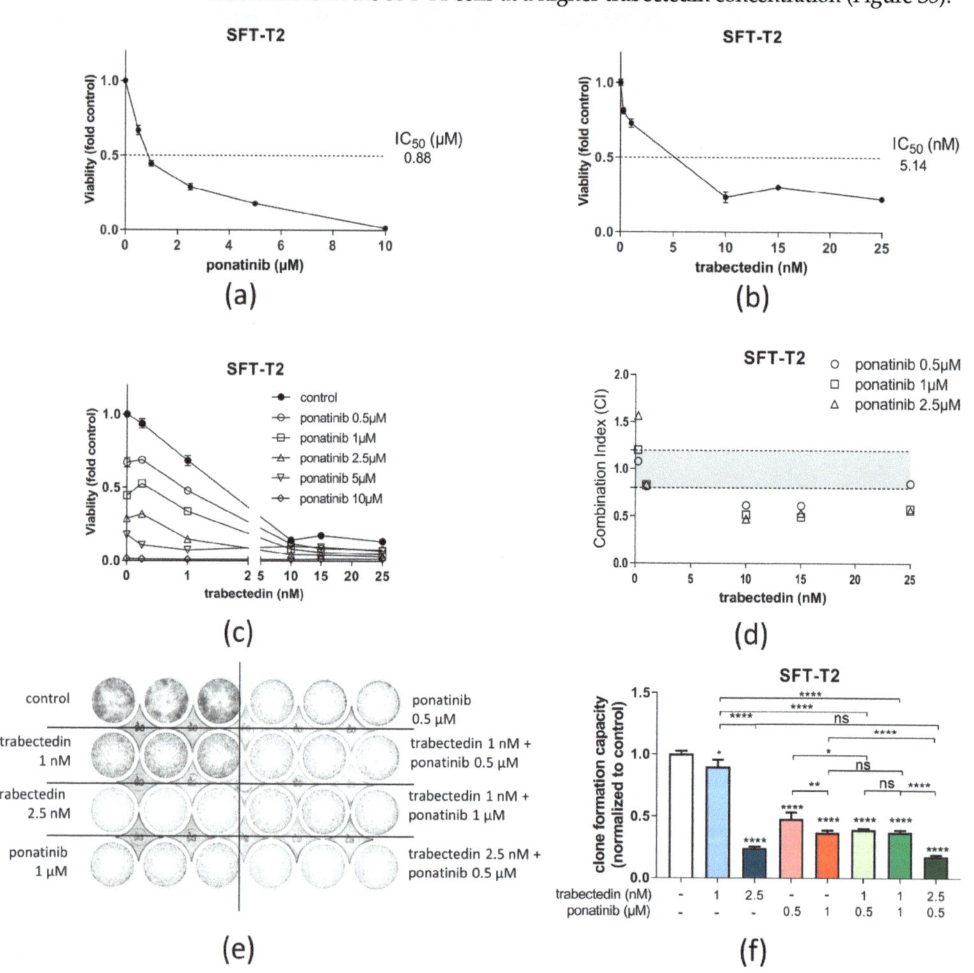

Figure 5. Ponatinib and trabectedin are active against SFT-T2 cell growth and synergize in vitro. Growth inhibitory effects of (**a**) ponatinib or (**b**) trabectedin as single agents in SFT-T2 cells assessed via MTT assay for 72 h of treatment. (**c**) Combination of ponatinib with trabectedin shows synergistic anticancer activity. (**d**) Corresponding CI values. CI < 0.9, synergism; CI = 0.9–1.2, additive effects; or CI > 1.2, antagonism. (**e**) Effects of ponatinib as well as trabectedin as single agents and in combination on SFT clone formation capacity over a treatment period of nine days. (**f**) Quantification of the clone formation assay. * $p < 0.05$, ** $p < 0.005$ and **** $p < 0.0001$.

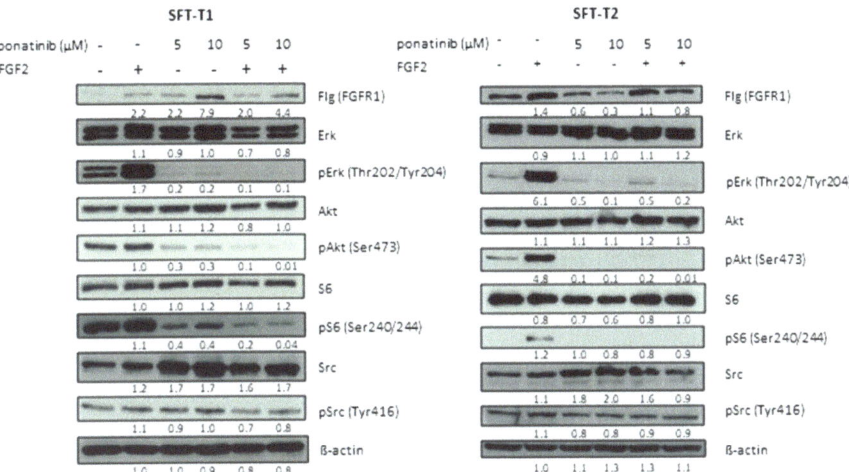

Figure 6. Ponatinib inhibits MAPK and PI3K/Akt signaling cascades in SFT-T1 and -T2 cells. Expression and phosphorylation levels of respective proteins after 1 h of treatment with indicated concentrations of ponatinib with or without stimulation with FGF2 for 15 min were analyzed by Western blotting. ß-actin served as loading control. Numbers below represent quantified signal intensities normalized to respective ß-actin relative to untreated control.

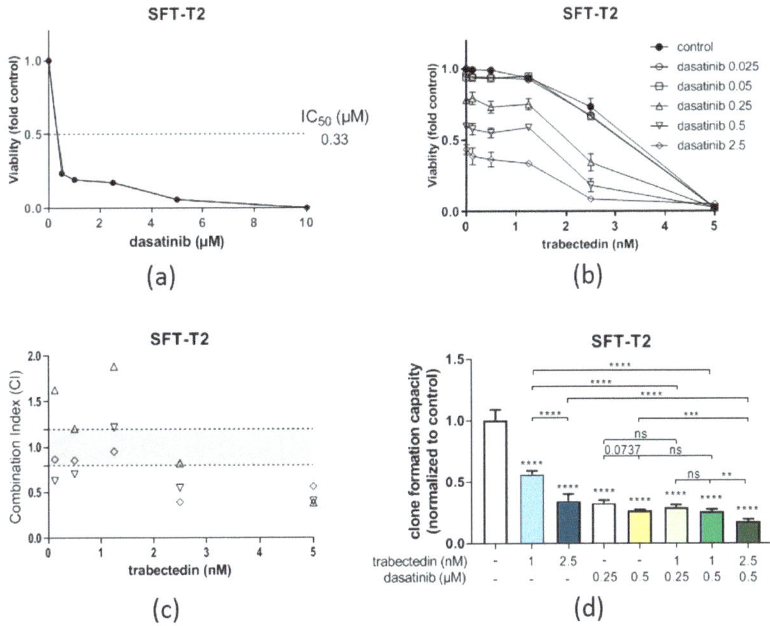

Figure 7. Dasatinib is active against SFT-T2 growth in vitro and enhances the anticancer activity of trabectedin. (**a**) Growth inhibitory effects of dasatinib as single agents in SFT-T2 cells assessed via MTT assay for 72 h of treatment. (**b**) Combination of dasatinib with trabectedin shows additive or even synergistic anticancer activity. (**c**) Corresponding CI values. CI < 0.9, synergism; CI = 0.9–1.2, additive effects; or CI > 1.2, antagonism. (**d**) Quantification of the effects of dasatinib as well as trabectedin as single agents and in combination on SFT-T2 clone formation capacity over a treatment period of nine days. ** $p < 0.005$, *** $p < 0.0005$, and **** $p < 0.0001$.

4. Discussion

SFT is a rare disease accounting for less than 5% of all pleural tumors [34]. Concerning prognostic biomarkers, even the most validated SFT prognosticators such as the Demicco classification [25] failed to estimate the outcome in SFT of the bone [35], and thus, an established staging and risk stratification is still missing for this rare malignancy. The standard therapy for local SFT presents radical surgery with or without radiotherapy [12,16,36]. At the disseminated stage, curative systemic treatment options are currently not available. Hence, SFT deserves attention regarding in-depth investigations of potential treatment approaches including drug repurposing strategies with the aim to improve the prognosis for this progressive disease. However, most of the explored systemic treatment approaches failed to significantly improve outcomes of SFT patients thus far [13–16,33,37]. Due to the low incidence numbers, most of the studies on SFT were of a retrospective nature and/or included only small numbers of SFT patients. Furthermore, the availability of data derived from primary SFT cell and xenograft models is even scarcer [16]. Thus, in the present study we investigated different innovative treatment approaches in two newly established fusion-positive, SFT-derived cell models to deliver preclinical data for potential further validation in the clinical setting.

Despite the low incidence numbers and other challenges of successful SFT in vitro cultivation, we were able to establish—from 14 surgical specimens with an initial SFT diagnosis—two permanent SFT cell models. Of note, the cell models presented here are—to the best of our knowledge—the first SFT patient-derived cell lines harboring the *NAB2-STAT6* fusion, proven at the mRNA and protein level. Regarding preclinical SFT models, the *NAB2-STAT6* fusion positivity was, thus far, only reported in case of two SFT patient-derived xenograft (PDX) models, out of which only one exhibited nuclear localization of STAT6 [37]. It should be mentioned here that, in our cell models, we always detected the expression of wild-type STAT6 together with the fusion protein, as also described by Robinson et al. [7], suggesting a potential cooperative function of the fusion with the wild-type proteins. However, at least two additional permanent SFT cell cultures from our collection had lost the transgene expression during in vitro propagation despite the fact that the origin from the *NAB2-STAT6* RNA-positive patient sample was proven by Short Tandem Repeat analysis. Interestingly, the loss of the fusion protein was described as a marker of so-called "dedifferentiated SFT" [38]. Whether comparable processes happened during the in vitro propagation of our *NAB2-STAT6*-negative cell models or cell clones with fusion protein loss were already present in the original tumor is a matter of ongoing investigations.

In the current study, we focused on systemic treatment options for the classical fusion-positive genotype of SFT. Hence, we employed our two fusion protein-expressing, novel SFT models to test an extended panel of chemo- and targeted therapeutics concerning their anti-SFT activity. Moreover, we performed molecular analyses for the most promising compounds regarding the interaction with FGFR signaling, due to the previously reported overexpression of FGFR1 in SFT [9,10]. Another important aspect of this work included the evaluation of the most effective therapy combinations in the preclinical setting. Furthermore, the already approved compound trabectedin for sarcoma patients was used in both investigated combination regimens. This aspect might support clinical testing of our drug combinations.

In line with the clinical behavior of SFT, our fusion-positive cell models were insensitive to most of the tested systemic treatments, as shown in Table 2. Among the relatively active compounds was trabectedin with IC_{50} values in the low nM range. Trabectedin is a marine alkaloid with DNA-interacting and immune-stimulatory activities [39] and was approved for treatment of advanced soft tissue sarcoma and recurrent ovarian carcinoma [40]. Trabectedin showed remarkable activity in translocation-related sarcomas by modulating the corresponding fusion oncogenes, as reviewed before [29,41]. As a member of the translocation-related soft tissue sarcomas, early retrospective studies suggested the efficacy of trabectedin also against SFT [19,42,43]. Indeed, promising results of trabectedin in treating the first two available preclinical SFT PDX models have been reported [37]. The strong cytotoxic potential of trabectedin against our two fusion protein-positive SFT models further suggests a clinical evaluation of the compound against progressing SFT. Moreover, it was shown that trabectedin modulates the tumor microenvironment by targeting tumor-promoting inflammatory cell compartments [29]. This, indeed, might enhance the clinical efficacy of trabectedin, since tumor-promoting inflammation is known to play an important role in malignant pleural disease including SFT, as also demonstrated by our group [12,44].

Despite the distinct activity of trabectedin in the PDX models, the outcome was not curative and tumors relapsed [37]. Consequently, we screened for promising combination approaches with clinically approved anticancer agents to enhance the—in part—promising clinical and preclinical activity of trabectedin. Other highly active compounds identified in our drug screen were ponatinib (a clinically approved multi-kinase inhibitor of FGFR, PDGFR, VEGFR, Abl kinase, and c-Kit), and dasatinib (inhibitor of Bcr-Abl, Src, and c-Kit). Both of these compounds were shown to exert antiangiogenic properties, another reported vulnerability of SFT [16]. Ponatinib seemed to be especially interesting, since one of its central targets are FGFRs, including FGFR1, which is frequently overexpressed in SFT [9,10]. Additionally, the FGFR inhibitor regorafenib displayed the highest growth inhibitor effects out of all tyrosine kinase inhibitors tested against a dedifferentiated SFT PDX model [45]. In addition, the multi-tyrosine kinase inhibitor pazopanib (VEGFRs, PDGFR, and cKit), that also has some modest efficacy against FGF signaling, was active as first-line therapy in metastatic SFT in a prospective study [46]. However, when it comes to the dedifferentiated subtype in the aforementioned PDX model, the more FGF-targeting therapy regorafinib proved to have the highest anti-SFT activity, and pazopanib was only moderately active [45].

Thus, one might assume ponatinib as the most promising systemic treatment option for SFT among the multi-tyrosine kinase inhibitors due to its (1) antiangiogenic, (2) fusion/translation, and (3) FGFR-targeting characteristics [5,16,30]. The latter was identified in our in vitro analyses to play a crucial role. Indeed, our novel SFT models were also highly responsive to FGF2-mediated signals by upregulating MAPK as well as PI3K/Akt FGFR downstream pathways. Interestingly, not only FGF2-induced, but also the basal activation of these pro-oncogenic signaling cascades was clearly downregulated in SFT-T1 and even completely blocked in SFT-T2 by exposure to ponatinib. As the used SFT cell culture medium was devoid of any FGF supplementation, these observations strongly suggest an autocrine FGFR-based growth and survival signaling loop by endogenously expressed ligands playing a central role in basal MAPK signal maintenance in SFT. This assumption is supported by early studies suggesting FGF2 expression as a diagnostic and prognostic marker in SFT [11]. Of note, it has to be mentioned that three effective drugs out of the drug efficacy screening were also targeting the FGF axis (ponatinib, nintedanib, PD173074—compare Table 2), again underlining a potential interesting and targetable role of FGFR in SFT.

Src as a target of ponatinib appears to be of minor importance in our SFT models since the expression and phosphorylation of Src were not changed by treatment with a multi-tyrosine kinase inhibitor. However, to further test the role of Src as a potential target in SFT, we used the Src, c-Kit, and Abl kinase inhibitor dasatinib. Dasatinib exerted substantial anti-SFT cell activity in our novel models. The mechanisms underlying the impact are enigmatic as neither c-Kit, Abl kinase, nor Src seem to be major players in the SFT malignant phenotype. Accordingly, the first clinical data about dasatinib activity against sarcoma growth including 25 SFT patients were disappointing [33]. Nevertheless, this poor clinical single-agent activity might be improved after combining dasatinib with trabectedin, a combination that was found to exert additive to synergistic anti-SFT growth activity in vitro in both novel SFT cell models.. These findings might also be an interesting issue for clinical follow-up studies. In general, it seems reasonable to combine trabectedin with another targeted therapy to achieve the best response rates in this treatment-resistant disease.

5. Conclusions

During the project, we were able to establish two patient-derived SFT cell lines. Both cell lines harbored the SFT-characteristic *NAB2-STAT6* fusion. This achievement facilitates in vitro drug testing in an orphan disease, where large, prospective, randomized, SFT-specific trials have been missing thus far. Our experiments provided evidence that fusion-positive SFT—as a generally treatment-resistant disease—is responsive to trabectedin in vitro. In addition, we were able to show that the tested SFT cell lines were strongly responsive to FGF2 stimulation, and basal MAPK/PI3K signaling was sensitive towards the FGFR inhibitor ponatinib. Furthermore, and of clinical relevance, we were able to propose synergistic treatment combinations, including trabectedin with the multi-tyrosine kinase inhibitors ponatinib and dasatinib. In particular, the combination of trabectedin with ponatinib might represent a promising treatment approach in this otherwise resistant disease, and is worth being validated in the clinical setting.

Supplementary Materials: The following supporting information can be downloaded at: https://www.mdpi.com/article/10.3390/cancers14225602/s1. Figure S1. Representative (immune-)histological pictures of negative control stainings with isotypes corresponding to the antibodies (a) CD34, (b) bcl-2, (c) CD99, (d) Ki67, (e) STAT6 of the two included tumors (SFT-T1 and SFT-T2) and an excluded fibromatosis patient (Fibr) compare Figure 2 of the main text. Figure S2. SFT-T1, SFT-T2, HCC827, and Met5a cells show stable in vitro growth. (a) Representative digital phase contrast images of respective cell lines after T0, 24, 48, and 72 h. (b) Quantification of pixel intensities of images shown in (a); n = 3. Figure S3. Representative in vitro dose-response curves of SFT-T1 (upper) and SFT-T2 (lower) cells after 72 h exposure to (a) cisplatin, (b) paclitaxel, (c) vincristine, (d) venetoclax, (e) obatoclax, (f) doxorubicin, (g) imatinib, (h) stattic, (i) trabectedin, (j) ponatinib, (k) nintedanib, (l) dasatinib, and (m) PD173074 determined by MTT assay. Figure S4. Ponatinib and trabectedin are active against low risk group SFT-T1 cell growth and synergize in vitro. (a) Combination of ponatinib with trabectedin shows synergistic anticancer activity. (b) Corresponding CI values. CI < 0.9, synergism; CI = 0.9–1.2, additive effects; or CI > 1.2 antagonism. (c) Quantification of effects of ponatinib and trabectedin as single agents and in combination on SFT-T1 clone formation capacity over a treatment period of nine days. Figure S5. Dasatinib is active against SFT-T1 growth in vitro and enhances the anticancer activity of trabectedin. Effects of trabectedin as well as dasatinib as single agents and in combination on SFT clone formation capacity over a treatment period of nine days. Figure S6. Original Western blots of (a) Figure 3b, (b) Figure 3d, and (c) Figure 6.

Author Contributions: Conceptualization, B.G., D.B. and W.B.; data curation, B.G., D.B. and W.B.; formal analysis, B.G., D.B., C.P. and L.M.; funding acquisition, B.G. and W.B.; investigation, B.G., D.B., C.P., L.M. and K.S.; methodology, B.G., D.B., C.P., L.M. and W.B.; project administration, B.G. and G.L.; resources, B.G., G.L., K.H. and W.B.; supervision, B.G., G.L., K.H. and W.B.; validation, B.G., D.B., C.P., L.M. and W.B.; visualization, B.G., D.B., C.P. and L.M.; writing—original draft, B.G. and D.B.; writing—review and editing, B.G., D.B., C.P., L.M., K.S., G.L., K.H. and W.B. All authors have read and agreed to the published version of the manuscript. All authors have approved the present work.

Funding: The whole work was financed by the Margaretha Hehberger Stiftung (#15129), Vienna, Austria.

Institutional Review Board Statement: The study was approved by the Medical University of Vienna Ethics Committee under the study title, "Etablierung einer Tumorbank für die molekulare Analyse von thorakalen Tumoren", EK Nr: 9004/2009. The whole study was conducted according to the Helsinki Declaration and the guidelines for good scientific practice.

Informed Consent Statement: Informed consent was obtained from all subjects involved in the study.

Data Availability Statement: Upon reasonable request, all data and materials are available from the first author.

Acknowledgments: The authors want to thank Jaqueline Blank, Lisa Wozelka, and Barbara Neudert for help with next-generation sequencing and immunohistochemistry. The authors express great appreciation to the technicians Petra Vician and Mirjana Stojanovic for their assistance, as well as to biomedical scientists Georg Schröckenfuchs and Jennifer Hsu for biobank cultivation.

Conflicts of Interest: The authors have no conflict of interest related to the presented work to declare.

References

1. Stiller, C.A.; Trama, A.; Serraino, D.; Rossi, S.; Navarro, C.; Chirlaque, M.D.; Casali, P.G.; Group, R.W. Descriptive epidemiology of sarcomas in Europe: Report from the RARECARE project. *Eur. J. Cancer.* **2013**, *49*, 684–695. [CrossRef] [PubMed]
2. England, D.M.; Hochholzer, L.; McCarthy, M.J. Localized benign and malignant fibrous tumors of the pleura. A clinicopathologic review of 223 cases. *Am. J. Surg. Pathol.* **1989**, *13*, 640–658. [CrossRef] [PubMed]
3. de Perrot, M.; Fischer, S.; Brundler, M.A.; Sekine, Y.; Keshavjee, S. Solitary fibrous tumors of the pleura. *Ann. Thorac. Surg.* **2002**, *74*, 285–293. [CrossRef] [PubMed]
4. Abu Arab, W. Solitary fibrous tumours of the pleura. *Eur. J. Cardiothorac. Surg.* **2012**, *41*, 587–597. [CrossRef]
5. Kallen, M.E.; Hornick, J.L. The 2020 WHO Classification: What's New in Soft Tissue Tumor Pathology? *Am. J. Surg. Pathol.* **2021**, *45*, e1–e23. [CrossRef]
6. Chmielecki, J.; Crago, A.M.; Rosenberg, M.; O'Connor, R.; Walker, S.R.; Ambrogio, L.; Auclair, D.; McKenna, A.; Heinrich, M.C.; Frank, D.A.; et al. Whole-exome sequencing identifies a recurrent NAB2-STAT6 fusion in solitary fibrous tumors. *Nat. Genet.* **2013**, *45*, 131–132. [CrossRef]
7. Robinson, D.R.; Wu, Y.M.; Kalyana-Sundaram, S.; Cao, X.; Lonigro, R.J.; Sung, Y.S.; Chen, C.L.; Zhang, L.; Wang, R.; Su, F.; et al. Identification of recurrent NAB2-STAT6 gene fusions in solitary fibrous tumor by integrative sequencing. *Nat. Genet.* **2013**, *45*, 180–185. [CrossRef]
8. Yoshida, A.; Tsuta, K.; Ohno, M.; Yoshida, M.; Narita, Y.; Kawai, A.; Asamura, H.; Kushima, R. STAT6 immunohistochemistry is helpful in the diagnosis of solitary fibrous tumors. *Am. J. Surg. Pathol.* **2014**, *38*, 552–559. [CrossRef]
9. Bertucci, F.; Bouvier-Labit, C.; Finetti, P.; Metellus, P.; Adelaide, J.; Mokhtari, K.; Figarella-Branger, D.; Decouvelaere, A.V.; Miquel, C.; Coindre, J.M.; et al. Gene expression profiling of solitary fibrous tumors. *PLoS ONE* **2013**, *8*, e64497. [CrossRef]
10. Hajdu, M.; Singer, S.; Maki, R.G.; Schwartz, G.K.; Keohan, M.L.; Antonescu, C.R. IGF2 over-expression in solitary fibrous tumours is independent of anatomical location and is related to loss of imprinting. *J. Pathol.* **2010**, *221*, 300–307. [CrossRef]
11. Sun, Y.; Naito, Z.; Ishiwata, T.; Maeda, S.; Sugisaki, Y.; Asano, G. Basic FGF and Ki-67 proteins useful for immunohistological diagnostic evaluations in malignant solitary fibrous tumor. *Pathol. Int.* **2003**, *53*, 284–290. [CrossRef]
12. Ghanim, B.; Hess, S.; Bertoglio, P.; Celik, A.; Bas, A.; Oberndorfer, F.; Melfi, F.; Mussi, A.; Klepetko, W.; Pirker, C.; et al. Intrathoracic solitary fibrous tumor—An international multicenter study on clinical outcome and novel circulating biomarkers. *Sci. Rep.* **2017**, *7*, 12557. [CrossRef] [PubMed]
13. DeVito, N.; Henderson, E.; Han, G.; Reed, D.; Bui, M.M.; Lavey, R.; Robinson, L.; Zager, J.S.; Gonzalez, R.J.; Sondak, V.K.; et al. Clinical Characteristics and Outcomes for Solitary Fibrous Tumor (SFT): A Single Center Experience. *PLoS ONE* **2015**, *10*, e0140362. [CrossRef] [PubMed]
14. Levard, A.; Derbel, O.; Meeus, P.; Ranchere, D.; Ray-Coquard, I.; Blay, J.Y.; Cassier, P.A. Outcome of patients with advanced solitary fibrous tumors: The Centre Leon Berard experience. *BMC Cancer* **2013**, *13*, 109. [CrossRef]
15. Schoffski, P.; Timmermans, I.; Hompes, D.; Stas, M.; Sinnaeve, F.; De Leyn, P.; Coosemans, W.; Van Raemdonck, D.; Hauben, E.; Sciot, R.; et al. Clinical Presentation, Natural History, and Therapeutic Approach in Patients with Solitary Fibrous Tumor: A Retrospective Analysis. *Sarcoma* **2020**, *2020*, 1385978. [CrossRef] [PubMed]
16. Martin-Broto, J.; Mondaza-Hernandez, J.L.; Moura, D.S.; Hindi, N. A Comprehensive Review on Solitary Fibrous Tumor: New Insights for New Horizons. *Cancers* **2021**, *13*, 2913. [CrossRef]
17. Park, M.S.; Ravi, V.; Conley, A.; Patel, S.R.; Trent, J.C.; Lev, D.C.; Lazar, A.J.; Wang, W.L.; Benjamin, R.S.; Araujo, D.M. The role of chemotherapy in advanced solitary fibrous tumors: A retrospective analysis. *Clin. Sarcoma Res.* **2013**, *3*, 1–7. [CrossRef]
18. Valentin, T.; Fournier, C.; Penel, N.; Bompas, E.; Chaigneau, L.; Isambert, N.; Chevreau, C. Sorafenib in patients with progressive malignant solitary fibrous tumors: A subgroup analysis from a phase II study of the French Sarcoma Group (GSF/GETO). *Investig. New Drugs* **2013**, *31*, 1626–1627. [CrossRef]
19. Khalifa, J.; Ouali, M.; Chaltiel, L.; Le Guellec, S.; Le Cesne, A.; Blay, J.Y.; Cousin, P.; Chaigneau, L.; Bompas, E.; Piperno-Neumann, S.; et al. Efficacy of trabectedin in malignant solitary fibrous tumors: A retrospective analysis from the French Sarcoma Group. *BMC Cancer* **2015**, *15*, 700. [CrossRef]
20. Hoda, M.A.; Mohamed, A.; Ghanim, B.; Filipits, M.; Hegedus, B.; Tamura, M.; Berta, J.; Kubista, B.; Dome, B.; Grusch, M.; et al. Temsirolimus inhibits malignant pleural mesothelioma growth in vitro and in vivo: Synergism with chemotherapy. *J. Thorac. Oncol.* **2011**, *6*, 852–863. [CrossRef]
21. Chou, T.C. Drug combination studies and their synergy quantification using the Chou-Talalay method. *Cancer Res.* **2010**, *70*, 440–446. [CrossRef] [PubMed]
22. Berger, W.; Elbling, L.; Micksche, M. Expression of the major vault protein LRP in human non-small-cell lung cancer cells: Activation by short-term exposure to antineoplastic drugs. *Int. J. Cancer* **2000**, *88*, 293–300. [CrossRef]
23. Berger, W.; Hauptmann, E.; Elbling, L.; Vetterlein, M.; Kokoschka, E.M.; Micksche, M. Possible role of the multidrug resistance-associated protein (MRP) in chemoresistance of human melanoma cells. *Int. J. Cancer* **1997**, *71*, 108–115. [CrossRef]
24. Guseva, N.V.; Tanas, M.R.; Stence, A.A.; Sompallae, R.; Schade, J.C.; Bossler, A.D.; Bellizzi, A.M.; Ma, D. The NAB2-STAT6 gene fusion in solitary fibrous tumor can be reliably detected by anchored multiplexed PCR for targeted next-generation sequencing. *Cancer Genet.* **2016**, *209*, 303–312. [CrossRef] [PubMed]

25. Demicco, E.G.; Wagner, M.J.; Maki, R.G.; Gupta, V.; Iofin, I.; Lazar, A.J.; Wang, W.L. Risk assessment in solitary fibrous tumors: Validation and refinement of a risk stratification model. *Mod. Pathol.* **2017**, *30*, 1433–1442. [CrossRef] [PubMed]

26. Park, H.K.; Yu, D.B.; Sung, M.; Oh, E.; Kim, M.; Song, J.Y.; Lee, M.S.; Jung, K.; Noh, K.W.; An, S.; et al. Molecular changes in solitary fibrous tumor progression. *J. Mol. Med.* **2019**, *97*, 1413–1425. [CrossRef]

27. Or, C.R.; Chang, Y.; Lin, W.C.; Lee, W.C.; Su, H.L.; Cheung, M.W.; Huang, C.P.; Ho, C.; Chang, C.C. Obatoclax, a Pan-BCL-2 Inhibitor, Targets Cyclin D1 for Degradation to Induce Antiproliferation in Human Colorectal Carcinoma Cells. *Int. J. Mol. Sci.* **2016**, *18*, 44. [CrossRef]

28. Steele, T.M.; Talbott, G.C.; Sam, A.; Tepper, C.G.; Ghosh, P.M.; Vinall, R.L. Obatoclax, a BH3 Mimetic, Enhances Cisplatin-Induced Apoptosis and Decreases the Clonogenicity of Muscle Invasive Bladder Cancer Cells via Mechanisms That Involve the Inhibition of Pro-Survival Molecules as Well as Cell Cycle Regulators. *Int. J. Mol. Sci.* **2019**, *20*, 1285. [CrossRef]

29. D'Incalci, M.; Badri, N.; Galmarini, C.M.; Allavena, P. Trabectedin, a drug acting on both cancer cells and the tumour microenvironment. *Br. J. Cancer* **2014**, *111*, 646–650. [CrossRef]

30. Musumeci, F.; Greco, C.; Grossi, G.; Molinari, A.; Schenone, S. Recent Studies on Ponatinib in Cancers Other Than Chronic Myeloid Leukemia. *Cancers* **2018**, *10*, 430. [CrossRef]

31. Schelch, K.; Hoda, M.A.; Klikovits, T.; Munzker, J.; Ghanim, B.; Wagner, C.; Garay, T.; Laszlo, V.; Setinek, U.; Dome, B.; et al. Fibroblast growth factor receptor inhibition is active against mesothelioma and synergizes with radio- and chemotherapy. *Am. J. Respir. Crit. Care Med.* **2014**, *190*, 763–772. [CrossRef] [PubMed]

32. Zeng, P.; Schmaier, A. Ponatinib and other CML Tyrosine Kinase Inhibitors in Thrombosis. *Int. J. Mol. Sci.* **2020**, *21*, 6556. [CrossRef] [PubMed]

33. Schuetze, S.M.; Bolejack, V.; Choy, E.; Ganjoo, K.N.; Staddon, A.P.; Chow, W.A.; Tawbi, H.A.; Samuels, B.L.; Patel, S.R.; von Mehren, M.; et al. Phase 2 study of dasatinib in patients with alveolar soft part sarcoma, chondrosarcoma, chordoma, epithelioid sarcoma, or solitary fibrous tumor. *Cancer* **2017**, *123*, 90–97. [CrossRef] [PubMed]

34. Jeon, H.W.; Kwon, S.S.; Kim, Y.D. Malignant solitary fibrous tumor of the pleura slowly growing over 17 years: Case report. *J. Cardiothorac. Surg.* **2014**, *9*, 113. [CrossRef] [PubMed]

35. Bianchi, G.; Lana, D.; Gambarotti, M.; Ferrari, C.; Sbaraglia, M.; Pedrini, E.; Pazzaglia, L.; Sangiorgi, L.; Bartolotti, I.; Dei Tos, A.P.; et al. Clinical, Histological, and Molecular Features of Solitary Fibrous Tumor of Bone: A Single Institution Retrospective Review. *Cancers* **2021**, *13*, 2470. [CrossRef]

36. Apra, C.; El Arbi, A.; Montero, A.S.; Parker, F.; Knafo, S. Spinal Solitary Fibrous Tumors: An Original Multicenter Series and Systematic Review of Presentation, Management, and Prognosis. *Cancers* **2022**, *14*, 2839. [CrossRef]

37. Stacchiotti, S.; Saponara, M.; Frapolli, R.; Tortoreto, M.; Cominetti, D.; Provenzano, S.; Negri, T.; Dagrada, G.P.; Gronchi, A.; Colombo, C.; et al. Patient-derived solitary fibrous tumour xenografts predict high sensitivity to doxorubicin/dacarbazine combination confirmed in the clinic and highlight the potential effectiveness of trabectedin or eribulin against this tumour. *Eur. J. Cancer* **2017**, *76*, 84–92. [CrossRef]

38. Dagrada, G.P.; Spagnuolo, R.D.; Mauro, V.; Tamborini, E.; Cesana, L.; Gronchi, A.; Stacchiotti, S.; Pierotti, M.A.; Negri, T.; Pilotti, S. Solitary fibrous tumors: Loss of chimeric protein expression and genomic instability mark dedifferentiation. *Mod. Pathol.* **2015**, *28*, 1074–1083. [CrossRef]

39. Germano, G.; Frapolli, R.; Belgiovine, C.; Anselmo, A.; Pesce, S.; Liguori, M.; Erba, E.; Uboldi, S.; Zucchetti, M.; Pasqualini, F.; et al. Role of macrophage targeting in the antitumor activity of trabectedin. *Cancer Cell* **2013**, *23*, 249–262. [CrossRef]

40. Evangelisti, G.; Barra, F.; D'Alessandro, G.; Tantari, M.; Stigliani, S.; Della Corte, L.; Bifulco, G.; Ferrero, S. Trabectedin for the therapy of ovarian cancer. *Drugs Today* **2020**, *56*, 669–688. [CrossRef]

41. Nakano, K.; Takahashi, S. Translocation-Related Sarcomas. *Int. J. Mol. Sci.* **2018**, *19*, 3784. [CrossRef] [PubMed]

42. Chaigneau, L.; Kalbacher, E.; Thiery-Vuillemin, A.; Fagnoni-Legat, C.; Isambert, N.; Aherfi, L.; Pauchot, J.; Delroeux, D.; Servagi-Vernat, S.; Mansi, L.; et al. Efficacy of trabectedin in metastatic solitary fibrous tumor. *Rare Tumors* **2011**, *3*, e29. [CrossRef] [PubMed]

43. Kobayashi, H.; Iwata, S.; Wakamatsu, T.; Hayakawa, K.; Yonemoto, T.; Wasa, J.; Oka, H.; Ueda, T.; Tanaka, S. Efficacy and safety of trabectedin for patients with unresectable and relapsed soft-tissue sarcoma in Japan: A Japanese Musculoskeletal Oncology Group study. *Cancer* **2020**, *126*, 1253–1263. [CrossRef] [PubMed]

44. Vogl, M.; Rosenmayr, A.; Bohanes, T.; Scheed, A.; Brndiar, M.; Stubenberger, E.; Ghanim, B. Biomarkers for Malignant Pleural Mesothelioma-A Novel View on Inflammation. *Cancers* **2021**, *13*, 658. [CrossRef]

45. Stacchiotti, S.; Tortoreto, M.; Baldi, G.G.; Grignani, G.; Toss, A.; Badalamenti, G.; Cominetti, D.; Morosi, C.; Dei Tos, A.P.; Festinese, F.; et al. Preclinical and clinical evidence of activity of pazopanib in solitary fibrous tumour. *Eur. J. Cancer* **2014**, *50*, 3021–3028. [CrossRef]

46. Maruzzo, M.; Martin-Liberal, J.; Messiou, C.; Miah, A.; Thway, K.; Alvarado, R.; Judson, I.; Benson, C. Pazopanib as first line treatment for solitary fibrous tumours: The Royal Marsden Hospital experience. *Clin. Sarcoma Res.* **2015**, *5*, 1–7. [CrossRef]

Review

Spinal Solitary Fibrous Tumors: An Original Multicenter Series and Systematic Review of Presentation, Management, and Prognosis

Caroline Apra [1,2,*], **Amira El Arbi** [3], **Anne-Sophie Montero** [1,2], **Fabrice Parker** [3,4] and **Steven Knafo** [3,4,*]

1 Sorbonne Université, 75013 Paris, France; anne-sophie.montero@aphp.fr
2 Neurosurgery Department, Pitie Salpêtrière Hospital, 75013 Paris, France
3 Neurosurgery Department, Bicêtre Hospital, 94270 Kremlin-Bicêtre, France; mira.elarbi@gmail.com (A.E.A.); fabrice.parker@aphp.fr (F.P.)
4 University Paris-Saclay, 91190 Gif-sur-Yvette, France
* Correspondence: caroline.apra@neurochirurgie.fr (C.A.); steven.knafo@aphp.fr (S.K.)

Simple Summary: Solitary fibrous tumors are rare benign or cancerous tumors that develop in all tissues, including close to the spinal cord. These cases are exceptional and we describe their presentation and outcome based on 31 published cases and 10 patients on whom we operated. The tumors can develop in any portion of the spine and cause back pain, associated with neurological deficits, such as compression of a nerve or the spinal cord, in 66% of patients. Surgical removal is the first step towards diagnosis and treatment, but complete removal could be achieved in only 70% of patients, due to bleeding or spinal cord invasion. Tumors were found to recur after a mean 5.8 years (1 to 25), without identified risk factors. However, in patients with subtotal removal, radiotherapy significantly improves the rate of recurrence. In total, spinal solitary fibrous tumors are treated by neurosurgeons on the front line but discussion in a multidisciplinary team will provide general treatments, especially radiotherapy after subtotal removal.

Abstract: All solitary fibrous tumors (SFT), now histologically diagnosed by a positive nuclear STAT6 immunostaining, represent less than 2% of soft tissue sarcomas, with spinal SFT constituting a maximum of 2% of them, making these tumors extremely rare. We provide an up-to-date overview of their diagnosis, treatment, and prognosis. We included 10 primary STAT6-positive SFT from our retrospective cohort and 31 from a systematic review. Spinal pain was the most common symptom, in 69% of patients, and the only one in 34%, followed by spinal cord compression in 41%, radicular compression, including pain or deficit, in 36%, and urinary dysfunction specifically in 18%. Preoperative diagnosis was never obtained. Gross total resection was achieved in 71%, in the absence of spinal cord invasion or excessive bleeding. Histologically, they were 35% grade I, 25% grade II, and 40% grade III. Recurrence was observed in 43% after a mean 5.8 years (1 to 25). No significant risk factor was identified, but adjuvant radiotherapy improved the recurrence-free survival after subtotal resection. In conclusion, spinal SFT must be treated by neurosurgeons as part of a multidisciplinary team. Owing to their close relationship with the spinal cord, radiotherapy should be considered when gross total resection cannot be achieved, to lower the risk of recurrence.

Keywords: spine; medulla; intramedullary; solitary fibrous tumor; hemangiopericytoma; neurosurgery; STAT6

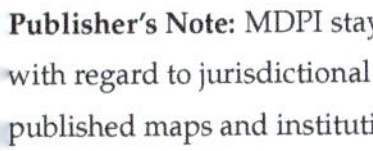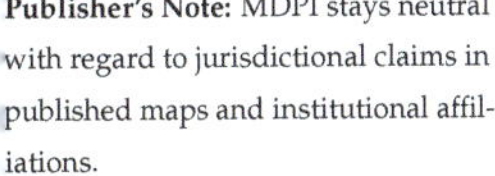 check for updates

Citation: Apra, C.; El Arbi, A.; Montero, A.-S.; Parker, F.; Knafo, S. Spinal Solitary Fibrous Tumors: An Original Multicenter Series and Systematic Review of Presentation, Management, and Prognosis. *Cancers* 2022, 14, 2839. https://doi.org/10.3390/cancers14122839

Academic Editor: David Wong

Received: 2 May 2022
Accepted: 2 June 2022
Published: 8 June 2022

Publisher's Note: MDPI stays neutral with regard to jurisdictional claims in published maps and institutional affiliations.

1. Introduction

In 2016, the World Health Organization (WHO) introduced the combined term "solitary fibrous tumor/hemangiopericytoma" for describing connective tissue tumors of the central nervous system with positive STAT6 nuclear immunostaining, which was replaced by "solitary fibrous tumors" (SFT) alone in 2021, to conform fully with soft tissue pathology nomenclature [1]. The grouping of these two entities, which were separated until then, is grounded in an overlapping histological description, associated with a shared genetic signature: the fusion of the NGFI-A-binding protein 2 (*NAB2*) and signal transducer and activator of transcription 6 (*STAT6*) genes due to an inversion at chromosome 12q13, which is a hallmark for all SFT, regardless of their localization, since its first description in 2013 [2]. SFT represent less than 2% of soft tissue sarcomas, with central nervous system SFT constituting 20% of SFT, with only one spinal SFT for every ten intracranial lesions, making these tumors extremely rare in clinical practice [2]. We performed a retrospective multicenter series of spinal SFT, focusing on STAT6-positive tumors only, and added a systematic review of all published cases. The aim of this review is to provide an up-to-date overview of the diagnosis, treatment, and prognosis of these rare tumors, with a discussion about individual clinical decision making.

2. Materials and Methods

2.1. Original Series

All patients with a histological diagnosis of spinal SFT who underwent surgery in the neurosurgical departments of the Bicêtre and Pitié-Salpêtrière hospitals in Paris, France between 1988 and 2020 were included. SFT diagnosis was confirmed by two expert neuropathologists with confirmed positive STAT6 nuclear immunostaining even for the cases dated before 2016. Medical records were reviewed for clinical and radiological presentation, histopathologic features, surgical treatment, postoperative therapies, and outcomes. Tumors were graded radiologically [3]: I, extradural type; II, intradural type; and III, intra-to extradural and paravertebral type. Extent of resection was estimated from the operative reports and postoperative MRI. Tumors were graded histologically according to the 2016 WHO classification [1]. Duration of follow-up was calculated as the duration from the date of surgery to the last outpatient department visit. Recurrence was defined as local tumor growth on MRI, whether symptomatic or not. Ethical approval was granted by the French Neurosurgical Society review board (IRB00011687; 2022/14).

2.2. Systematic Review

The literature review was performed according to the PRISMA checklist [4]. The database (PubMed) was searched for the combination of terms "spine", "spinal", "medulla", "medullary", "solitary fibrous tumor", "hemangiopericytoma" in December 2021. Eligible articles reported at least one case of spinal SFT. Exclusion criteria were the absence of STAT6-positive nuclear immunostaining, spinal metastatic localizations, and the absence of clinical data. Clinical, radiological, and histological data, including perioperative descriptions of the tumors, were collected.

2.3. Statistical Analysis

All statistical analyses were performed using Microsoft Excel (version 2202). For analyzing risk factors for recurrence, exact Fisher test was performed due to the small size of the population. Schematic figures were created with http://www.BioRender.com (accessed on 10 April 2022).

3. Results

3.1. Patients

We included 10 cases from our own retrospective cohort and retrieved 31 cases from the systematic review, as detailed in the Supplementary Flow Chart (Supplementary Figure S1) [5–20]. All were primary SFT with a STAT6-positive pathological diagnosis. Most information was available from all cases, as detailed in Supplementary Table S1. Results are given as: all patients [review; cohort]. There were slightly more women (51%, 55%/40%, sex ratio 1.05) and the mean age at diagnosis was 46 (45/47), ranging from 10 to 81 (10–81/32–71). All results are detailed in Table 1.

Table 1. Description of characteristics and prognosis, for the population presenting a spinal solitary fibrous tumor, as diagnosed with positive STAT6 nuclear staining, for our series (n = 10) and a systematic review of the literature (n = 31).

Characteristics		Our Series (n = 10)	Literature Review (n = 31)	Total (n = 41)
Population				
Sex	M	6 (60%)	14 (45%)	**20 (49%)**
	F	4 (40%)	17 (55%)	**21 (51%)**
Age, mean ± IC95		47 ± 8	45 ± 3	**46 ± 4**
Clinical and radiological presentations				
Spinal pain		5 (50%)	22 (76%)	**27 (69%)**
Radicular compression		6 (60%)	8 (28%)	**14 (36%)**
Spinal cord compression		7 (70%)	9 (31%)	**16 (41%)**
Urinary dysfunction		4 (40%)	3 (10%)	**7 (18%)**
Motor dysfunction		5 (50%)	9 (29%)	**14 (34%)**
Sensory dysfunction		4 (40%)	6 (19%)	**10 (24%)**
Duration of symptoms (mo)		10 ± 6	20 ± 12	**17 ± 9**
Tumor localization	Cervical	5 (50%)	7 (23%)	**12 (29%)**
	Thoracic	4 (40%)	17 (54%)	**21 (51%)**
	Lumbar	1 (10%)	7 (23%)	**8 (20%)**
Tumor type	I extradural	0 (0%)	2 (14%)	**2 (9%)**
	II intradural	7 (78%)	8 (57%)	**15 (65%)**
	III extra- and intradural	2 (22%)	4 (29%)	**6 (26%)**
Surgical and histological findings				
Complete resection		7 (70%)	22 (71%)	**29 (71%)**
Purely extramedullary tumor during surgery		5 (50%)	19 (68%)	**24 (63%)**
Histological grading	1	1 (11%)	6 (55%)	**7 (35%)**
	2	5 (56%)	0 (0%)	**5 (25%)**
	3	3 (33%)	5 (45%)	**8 (40%)**
Post-operative management and outcome				
Primary adjuvant treatment	None	6 (60%)	14 (77%)	**30 (73%)**
	Radiotherapy	4 (40%)	7 (23%)	**11 (27%)**
Documented recurrence		4 (40%)	11 (32%)	**15 (37%)**
Time to first recurrence (mo)		128 ± 116	49 ± 42	**70 ± 47**

3.2. Clinical Presentation

Spinal pain was the most common symptom, found in 27 patients (69% [76%; 50%]), and as an isolated symptom in 14 (34% [45%; 0%]). Radicular symptoms, including pain or deficit, were present in 14 patients (36% [28%; 60%]), whereas spinal cord compression, including increased reflexes or sensory–motor deficit, was present in 16 patients (41% [31%; 70%]), and urinary dysfunction was reported in seven patients (18%, [10%; 40%]). Symptoms typically worsened gradually, with a mean duration of clinical symptoms before surgery of 18 months (20/10), ranging from less than 1 month to 11 years. For patients with isolated spine pain, the mean duration of symptoms before surgery was 15 months.

3.3. Initial Radiological Findings

All patients underwent a preoperative spine MRI with contrast. Although T1 and T2 MRI aspects varied, all tumors showed marked homogeneous or heterogeneous enhancement by gadolinium. No calcification or acute intratumoral hemorrhage was observed. CT scan was not systematically available on retrieval, although it was likely performed in clinical practice before surgery. In a few cases, scalloping was observed in extracanalar tumors, but no exostosis. In our cohort, three patients had a preoperative angiography without embolization. Tumors were cervical in 12 patients (29% [23%; 50%]), thoracic in 21 patients (51% [54%; 40%]), and lumbar in eight patients (20% [23%; 10%]), which is proportional to the length of each segment, cervical spine vertebras constituting 30%, thoracic 50%, and lumbo-sacral 20% of the whole spine. Lesions involved one to two vertebrae in most patients (93% [97%; 80%]). Spinal SFT were type I in two patients (9% [14%; 0%]), II in 15 patients (65% [57%; 78%]), and III in six patients (26% [29%; 22%]). At least four [2; 2] extracanalar lesions showed extension in the foramina, giving them a dumbbell aspect.

3.4. Operative Findings

All patients underwent first-line surgery, through an isolated posterior approach with laminectomy, or associated with an anterior approach, an arthrodesis or a thoracoscopy in one case, depending on the tumor extension. Gross total resection was achieved in 29 patients (71% [71%; 70%]). The reasons for subtotal resection included spinal cord invasion and excessive bleeding. The use of neuro-monitoring was mentioned only once, after recurrence. Operatively, tumors were identified as being purely extramedullary in patients (63% [68%; 50%]) with or without description of dural and pial invasion, whereas other cases invaded the medulla. No intraoperative complication other than bleeding was noted, either in the literature or in our series. A video showing perioperative observations is available as a Supplementary File.

3.5. Histological Findings

All tumors were diagnosed as spinal SFT with positive STAT6 nuclear immunostaining. Seven [6; 1] were classified as grade I, five [0; 5] as grade II, and eight [5; 3] grade III. The mitoses count ranged from 0 to 15 per 10 high-power fields, and Ki67 from 0% to 15%. Other reported immunostainings include CD34, vimentin, proteinS100, EMA, and SMA, but these are not reliable in SFT diagnosis. No histological evidence of medullary invasion was described.

3.6. Adjuvant Treatment and Outcome

No complication, either infection, cerebrospinal fluid leakage, neurological worsening, or death, was reported after surgery. Eleven (27 [23%; 40%]) patients received immediate postoperative adjuvant treatments, including 11 radiotherapy and one neoadjuvant chemotherapy. The rationale for performing these treatments was not systematic and not explicit but these patients included one case of grade III SFT and four cases of subtotal resection. Available follow-up ranged from 12 months to 30 years for 35 [26; 9] patients. Recurrence was observed in 15 patients (43% [42%; 44%]), after a mean 5.8 years, ranging from 1 to 25 years, as detailed in Figure 1. No significant risk factor for recurrence could be identified, but there was a tendency to recur for patients with incomplete surgery or no adjuvant radiotherapy (Table 2). Survival analyses show that adjuvant radiotherapy significantly improves the recurrence-free survival in patients with subtotal resection but has no effect in patients with gross total resection (Figure 1). Repeated recurrences were observed in some cases, but data were scarce. After recurrence, nine (75% [63%; 100%]) patients underwent a second surgery, with a combination of radiotherapy, carbon-ion radiotherapy, or proton therapy, and one was treated with Pazopanib for progressive disease after third recurrence. There were no systematic data about metastasizing.

Table 2. Risk factors for recurrence in patients with a minimum 12 months of follow-up ($n = 35$), *p*-value for exact Fisher test. WHO: World Health Organization.

Risk Factor for Recurrence	Recurrence ($n = 15$)	No Recurrence ($n = 20$)	*p*-Value
Intramedullary component	36% ($n = 14$)	26% ($n = 29$)	0.70
Subtotal resection	53% ($n = 15$)	20% ($n = 20$)	0.07
WHO grade 3	40% ($n = 5$)	50% ($n = 10$)	1
Adjuvant radiotherapy	13% ($n = 15$)	40% ($n = 20$)	0.13

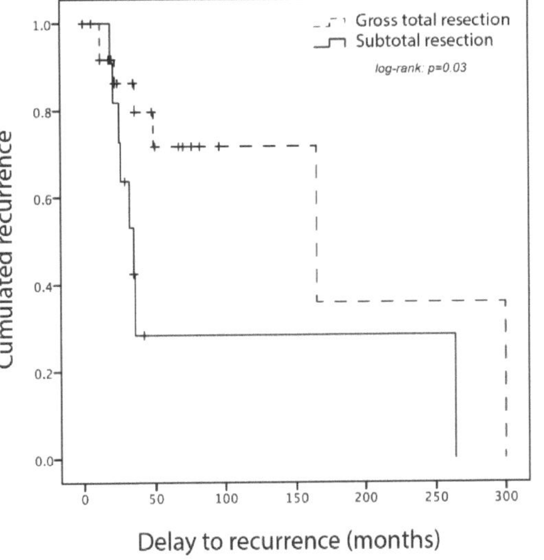

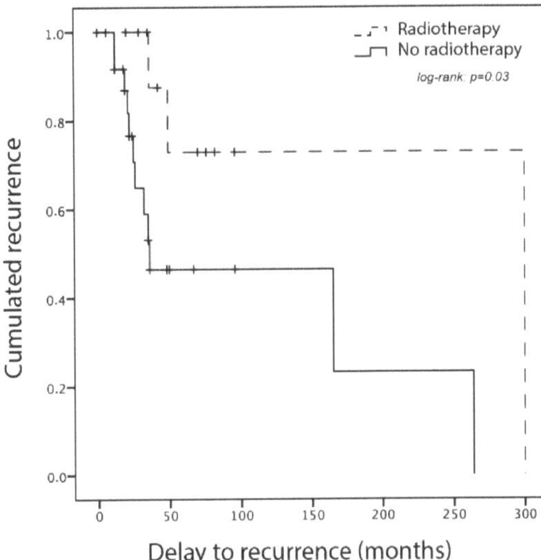

Figure 1. Recurrence-free survival (Kaplan–Meier curve) for all patients with a spinal solitary fibrous tumor and at least 12 months of follow-up, based on our series and literature systematic review (n = 35). Each + accounts for a patient death or end of follow-up. Left: for patients with gross total resection compared to subtotal resection. Right: for patients with adjuvant radiotherapy compared to no radiotherapy. Radiotherapy significantly improved recurrence-free survival in patients with subtotal resection.

4. Discussion

4.1. Clinical Approach to Spinal SFT Management Based on Data and Experience

Spinal SFT are extremely rare tumors, and patients usually present with nonspecific symptoms. Most neurosurgeons will probably operate zero to a few spinal SFT in their career, without even suspecting it before histological results. The clinical and radiological features of spinal SFT are summarized in Figure 2. The first step toward diagnosis is to perform a spine MRI, which will allow the diagnosis of a spinal tumor. Although some features have been described to identify spinal SFT, they are not diagnosed correctly on MRI, because of their rarity, lack of specific features, and variety of radiological presentations [21]. Depending on the radiological type of the lesion, they are misdiagnosed as schwannomas, meningiomas, hemangioblastomas, metastases from solid cancers, ependymomas, and osteosarcomas [3,6,8,13,18–20]. Even in cases when a biopsy was performed before surgical resection, in the literature review, the diagnosis was not obtained correctly [18,19]. MRI-based classification [3] has significant limitations in terms of surgical planning, since spinal cord invasion cannot be assessed reliably. As a result, some authors have proposed to classify spinal SFT as vertebral, paravertebral, spinal cord, or mixed, to allow better anatomical understanding [6].

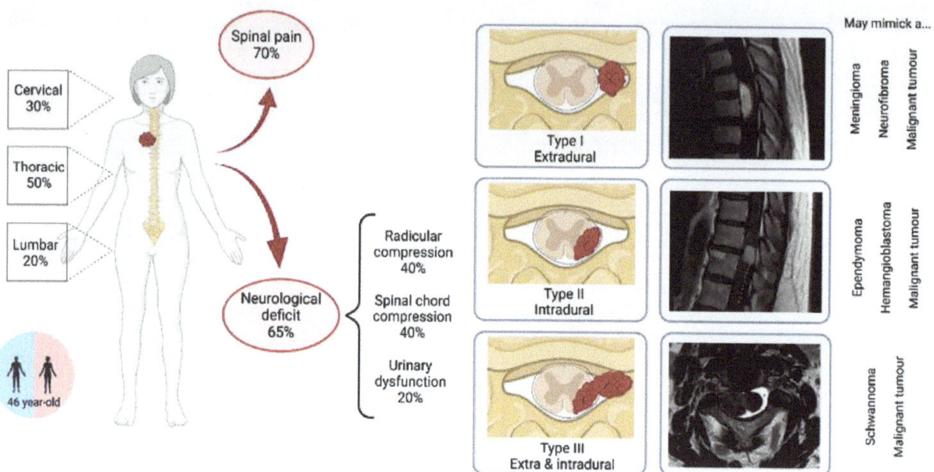

Figure 2. Graphical summary of the characteristics and clinico-radiological presentation of patients with spinal solitary fibrous tumors, including radiological types of SFT on MRI.

Nevertheless, whether SFT is suspected or not, surgery remains the first-line treatment option (Figure 3). A preoperative CT scan is recommended to assess bone invasion in all cases of spinal tumor diagnosis. The individual decision for arthrodesis is based on tumoral and surgical criteria, including articular process damage or destabilization due to an exceptionally large posterior laminectomy. Spinal angiography is useful for foraminal/anterior tumors located between T8 and L1 to identify the artery of Adamkiewicz, whose lesion can cause definitive paraplegia due to the interruption of the anterior spinal blood supply [22]. Preoperative embolization should be discussed every time spinal SFT is suspected, especially for large tumors or when percutaneous embolization is feasible and has proven to be useful in selected cases [23,24]. Preoperative neurological electrical assessment will rarely have an impact on the surgical decision making, except in paucisymptomatic patients, balancing in favor of surgery when a neurological impact arises. Perioperative neuro-monitoring is variably available in different hospitals but could be considered whenever intramedullary invasion is suspected. In addition, perioperative ultrasound may be useful in specific cases, when medullar invasion is suspected or to achieve recurrence removal, this technique usually confirming the surgeon's own microsurgical observation. There is evidence that 5-amino-levulinic acid induces fluorescence in spinal SFT, as in several other tumor types, which could help to identify the limits of invasive tumors, but its clinical usefulness needs to be proven [25]. In our experience, perioperative frozen histological analysis is seldom conclusive, but may rule out other diagnoses. As for any spinal tumor, the goals of surgery are to decompress the neurological structures and safely achieve resection, if possible complete [26]. The patient should be informed of the possible subtotal surgery and need for second-step surgery or adjuvant treatment.

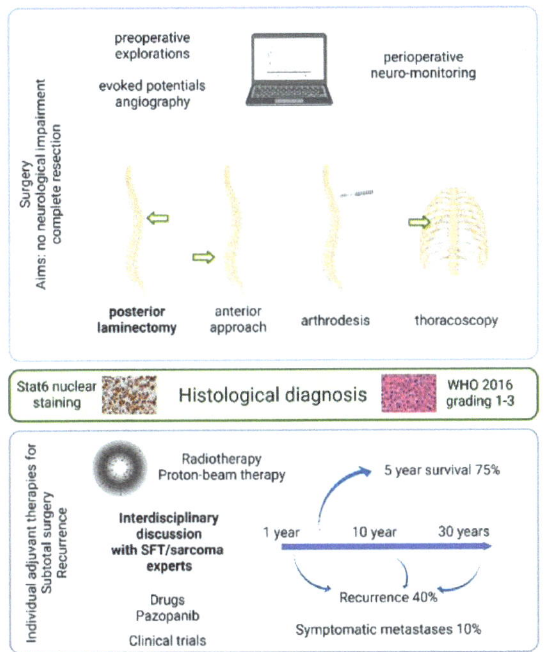

Figure 3. Graphical summary of the surgery, diagnosis, adjuvant treatments, and outcome of patients with spinal solitary fibrous tumors.

4.2. Postoperative Decision Making in Spinal SFT Treatment

Two main questions will arise at this point: First, what information can be reliably given to the patient concerning the tumor recurrence? Second, should any additional treatment be performed after surgery? Spinal SFT are extremely rare tumors. However, they are part of the SFT spectrum, which includes more common locations, including the rare intracranial SFT, and the more frequent pleural SFT [2,27]. All these tumors share the same genomic [2] and transcriptomic [28] identity, although they develop in different organs. Discussing these cases in multidisciplinary meetings with oncologists and surgeons aware of their histological rather than anatomic specificities can be of great help.

One factor that the patient must be aware of is the need for long-term follow-up, with recurrences happening up to 25 years after the initial surgery. Recurrences in meningeal SFT will happen in at least 37% of cases, after a mean 4.7 years, and symptomatic metastases in 10% of cases [26]. In spinal SFT, recurrences happen in 43% of patients after a mean 5.8 years and metastases in 11–25% [3,6]. Radiological or clinical follow-up is usually performed from every 3–6 months in the first few years after surgery to every 5 years life-long in the absence of any event. There is no indication to screen for asymptomatic metastases. However, it may be relevant to keep in mind that other meningeal localizations may occur, since up to two thirds of metastases actually are secondary intracranial or spinal SFT [3,12], and, in our experience, carcinomatous meningitis can also develop. Overall, the 5-year survival rate ranges between 76% and 93% [3].

Prognostic risk factors for recurrence that could help to decide about adjuvant treatment are controversial throughout the SFT literature. Recurrences in SFT in general are more likely to happen in tumors with a high diameter (superior to 6 cm), histological grade, necrosis, high mitotic rate, subtotal resection, absence of postoperative radiotherapy, and some localizations, including central nervous system [27,29,30]. From a molecular point of view, some types of NAB2–STAT6 fusions be associated with a worse prognosis, though not systematically [31]. Screening for the fusion type is not performed routinely and these results are not significant enough to make it necessary in clinical practice. Reviews that focus on spinal SFT, including us, failed to confirm any of these prognostic factors [3], except for subtotal resection in one study, which was significantly associated with a shorter recurrence-free survival and overall survival [6]. This review also identified WHO grades II-III as a risk factor for earlier recurrence but not survival, which could be associated with the progression of residual grade I SFT towards grade III [26].

Whether and when to perform adjuvant radiotherapy is still a matter of debate. There is evidence that adjuvant radiotherapy for both extrameningeal and meningeal SFT may improve local control [27,32–34]. However, adjuvant radiotherapy does not prevent the development of neuroaxis or peripheral metastases [34]. Compared to extrameningeal SFT, intradural lesions more often lead to subtotal resection, which is the main risk factor for local recurrence. Therefore, adjuvant radiotherapy could be offered after subtotal resection to delay local recurrence, keeping in mind that no effect on survival has been proven [3,6,34]. Moreover, radiation myelopathy, although rare, could significantly alter the quality of life of patients with a long survival. Stereotactic radiosurgery has been used for intracranial SFT, but its use in spinal SFT is sporadic and no conclusion can be drawn. Conventional chemotherapy gives a poor clinical benefit, and anti-angiogenic treatments are the most promising option [35–37], used in one patient in our series, as a third-line option.

4.3. Specificities of Spinal SFT

Although there is no controversy about the common molecular identity of SFT in all localizations since the description of NAB2–STAT6 fusion [2,28], spinal SFT is an ambiguous concept based on anatomy. They are usually considered meningeal SFT because they cause spinal cord compression and therefore neurological deficits. However, as illustrated by the wide variety of radiological and operative findings, it is not clear where these tumors arise from. Indeed, spinal SFT may well arise from the intradural space, from the vertebra [19,20], or from the pleura. Perioperative findings support the fact that these fibroblastic tumors arise from different layers, with some tumors clearly extramedullary [11], even extradural, whereas others present obvious signs of pial, nerve roots, or even spinal cord invasion [6,12,15]. Whether this variability is a sign of tumor aggressiveness or site of origin is not clear and no histological description of medullary invasion has been reported until now, whereas brain invasion has been reported in intracranial SFT, as in meningiomas [1].

This anatomical ambiguity correlates with the fact that the cell of origin of SFT is not determined: although it was advocated that meningeal SFT arise from a specific prostaglandin-D2-synthase-positive cell type, as with meningiomas [38,39], there is also molecular evidence that meningeal SFT probably share the same mesenchymal origin as all SFT [28]. This encourages us to treat spinal SFT as nonspecific to the central nervous system, questioning the fact that current clinical trials for SFT exclude patients with meningeal tumors.

5. Conclusions

Spinal SFT are extremely rare and versatile fibroblastic neoplasms with a high propensity to recur, representing a diagnostic and therapeutic challenge. Since clinical and radiological presentation does not usually allow preoperative diagnosis, surgery remains essential to achieve both diagnosis and neurological decompression. As spinal SFT are unequivocally part of the SFT spectrum in terms of molecular identity, they should be treated as such by a multidisciplinary team rather neurosurgeons alone. However, spinal SFT present specificities owing to its close relationship with the spinal cord. In particular, it seems that radiotherapy should be considered whenever gross total resection cannot be achieved due to spinal cord pial invasion given the significant rate of recurrence. Future developments of targeted therapies or neurologically sparing radiation protocols may help to control these tumors without damaging the surrounding neurological structures.

Supplementary Materials: The following supporting information can be downloaded at: https://www.mdpi.com/article/10.3390/cancers14122839/s1, Figure S1: Flow chart of the literature review for systematic review according to the PRSIMA guidelines, Table S1: Detailed anonymous data retrieval from our series and the systematic literature review for STAT6-positive spinal SFT.

Author Contributions: Conceptualization, C.A. and S.K.; methodology, C.A., F.P. and S.K.; formal analysis, C.A. and S.K.; investigation, C.A., A.E.A., A.-S.M. and S.K.; data curation, A.E.A. and A.-S.M.; writing—original draft preparation, C.A.; writing—review and editing, S.K.; supervision, F.P. and S.K. All authors have read and agreed to the published version of the manuscript.

Funding: This research received no external funding.

Institutional Review Board Statement: The study was conducted in accordance with the Declaration of Helsinki, and approved by the Institutional Review Board of the French Neurosurgical Society, College de Neurochirurgie (protocol code IRB00011687, 2022/14).

Informed Consent Statement: Informed consent was obtained from all subjects involved in the study.

Data Availability Statement: Data supporting reported results can be found in Supplementary Table S1.

Acknowledgments: The authors thank the neuropathology departments of hospitals Bicêtre (Adam) and Pitié-Salpêtrière (Mokhtari) for data retrieval and immunostaining confirmation.

Conflicts of Interest: The authors declare no conflict of interest.

References

1. Louis, D.N.; Perry, A.; Wesseling, P.; Brat, D.J.; Cree, I.A.; Figarella-Branger, D.; Hawkins, C.; Ng, H.K.; Pfister, S.M.; Reifenberger, G.; et al. The 2021 WHO Classification of Tumors of the Central Nervous System: A Summary. *Neuro Oncol.* **2021**, *23*, 1231–1251. [CrossRef] [PubMed]
2. Robinson, D.R.; Wu, Y.-M.; Kalyana-Sundaram, S.; Cao, X.; Lonigro, R.J.; Sung, Y.-S.; Chen, C.-L.; Zhang, L.; Wang, R.; Su, F.; et al. Identification of Recurrent NAB2-STAT6 Gene Fusions in Solitary Fibrous Tumor by Integrative Sequencing. *Nat. Genet.* **2013**, *45*, 180–185. [CrossRef] [PubMed]
3. Liu, H.; Yang, A.; Chen, N.; Yang, J.; Qiu, X.; Zhang, J. Hemangiopericytomas in the Spine: Clinical Features, Classification, Treatment, and Long-Term Follow-up in 26 Patients. *Neurosurgery* **2013**, *72*, 16–24; discussion 24 . [CrossRef] [PubMed]
4. Page, M.J.; McKenzie, J.E.; Bossuyt, P.M.; Boutron, I.; Hoffmann, T.C.; Mulrow, C.D.; Shamseer, L.; Tetzlaff, J.M.; Akl, E.A.; Brennan, S.E.; et al. The PRISMA 2020 Statement: An Updated Guideline for Reporting Systematic Reviews. *BMJ* **2021**, *372*, n71. [CrossRef]
5. Ando, M.; Kobayashi, H.; Shinozaki-Ushiku, A.; Chikuda, H.; Matsubayashi, Y.; Yoshida, M.; Saito, Y.; Kohsaka, S.; Oda, K.; Miyagawa, K.; et al. Spinal Solitary Fibrous Tumor of the Neck: Next-Generation Sequencing-Based Analysis of Genomic Aberrations. *Auris Nasus Larynx* **2020**, *47*, 1058–1063. [CrossRef]
6. Wang, J.; Zhao, K.; Han, L.; Jiao, L.; Liu, W.; Xu, Y.; Niu, H.; Ke, C.; Shu, K.; Lei, T. Solitary Fibrous Tumor/Hemangiopericytoma of Spinal Cord: A Retrospective Single-Center Study of 16 Cases. *World Neurosurg.* **2019**, *123*, e629–e638. [CrossRef]
7. Wang, L.; Yu, J.; Shu, D.; Huang, B.; Wang, Y.; Zhang, L. Primary Endodermal Hemangiopericytoma/Solitary Fibrous Tumor of the Cervical Spine: A Case Report and Literature Review. *BMC Surg.* **2021**, *21*, 405. [CrossRef]
8. Su, H.-Y.; Tsai, T.-H.; Yang, S.-F.; Lee, J.-Y. Dumbbell-Shaped Solitary Fibrous Tumor of Thoracic Spine. *Kaohsiung J. Med. Sci.* **2019**, *35*, 517–518. [CrossRef]
9. Flores-Justa, A.; López-García, E.; García-Allut, A.; Reyes-Santías, R.M. Solitary Fibrous Tumour/Haemangiopericytoma of the Spinal Cord. *Neurocirugia (Astur. Engl. E)* **2018**, *29*, 309–313. [CrossRef]
10. Mansilla Fernández, B.; Román de Aragón, M.; Paz Solís, J.F.; García Feijoo, P.; Roda Frade, J.; Regojo Zapata, M.R. Solitary Fibrous Tumor: A Clinical Case. *Neurocirugia (Astur. Engl. Ed.)* **2019**, *30*, 33–37. [CrossRef]
11. Olmsted, Z.T.; Tabor, J.; Doron, O.; Hosseini, H.; Schneider, D.; Green, R.; Wahl, S.J.; Sciubba, D.M.; D'Amico, R.S. Intradural Extramedullary Solitary Fibrous Tumor of the Thoracic Spinal Cord. *Cureus* **2021**, *13*, e18613. [CrossRef] [PubMed]
12. Albert, G.W.; Gokden, M. Solitary Fibrous Tumors of the Spine: A Pediatric Case Report with a Comprehensive Review of the Literature. *J. Neurosurg. Pediatr.* **2017**, *19*, 339–348. [CrossRef] [PubMed]
13. Dauleac, C.; Vasiljevic, A.; Berhouma, M. How to Differentiate Spinal Cord Hemangiopericytoma from Common Spinal Cord Tumor? *Neurochirurgie* **2020**, *66*, 53–55. [CrossRef] [PubMed]
14. Murata, K.; Endo, K.; Aihara, T.; Matsuoka, Y.; Nishimura, H.; Suzuki, H.; Sawaji, Y.; Yamamoto, K.; Fukami, S.; Tanigawa, M.; et al. Salvage Carbon Ion Radiotherapy for Recurrent Solitary Fibrous Tumor: A Case Report and Literature Review. *J. Orthop. Surg. (Hong Kong)* **2020**, *28*, 2309499019896099. [CrossRef] [PubMed]
15. Wei, D.; Ma, M.; Li, H. Invasive Solitary Fibrous Tumor/Hemangiopericytoma of the Filum Terminale. *World Neurosurg.* **2020**, *139*, 318–321. [CrossRef]

16. Shukla, P.; Gulwani, H.V.; Kaur, S.; Shanmugasundaram, D. Reappraisal of Morphological and Immunohistochemical Spectrum of Intracranial and Spinal Solitary Fibrous Tumors/Hemangiopericytomas with Impact on Long-Term Follow-Up. *Indian J. Cancer* **2018**, *55*, 214–221. [CrossRef]

17. Yao, Z.-G.; Wu, H.-B.; Hao, Y.-H.; Wang, X.-F.; Ma, G.-Z.; Li, J.; Li, J.-F.; Lin, C.-H.; Zhong, X.-M.; Wang, Z.; et al. Papillary Solitary Fibrous Tumor/Hemangiopericytoma: An Uncommon Morphological Form With NAB2-STAT6 Gene Fusion. *J. Neuropathol. Exp. Neurol.* **2019**, *78*, 685–693. [CrossRef]

18. Zhang, Y.-W.; Xiao, Q.; Zeng, J.-H.; Deng, L. Solitary Fibrous Tumor of the Lumbar Spine Resembling Schwannoma: A Case Report and Review of the Literature. *World Neurosurg.* **2019**, *124*, 121–124. [CrossRef]

19. Oike, N.; Kawashima, H.; Ogose, A.; Hotta, T.; Hirano, T.; Ariizumi, T.; Yamagishi, T.; Umezu, H.; Inagawa, S.; Endo, N. A Malignant Solitary Fibrous Tumour Arising from the First Lumbar Vertebra and Mimicking an Osteosarcoma: A Case Report. *World J. Surg. Oncol.* **2017**, *15*, 100. [CrossRef]

20. Farooq, Z.; Badar, Z.; Zaccarini, D.; Tavernier, F.B.; Mohamed, A.; Mangla, R. Recurrent Solitary Fibrous Tumor of Lumbar Spine with Vertebral Body Involvement: Imaging Features and Differential Diagnosis with Report of a Case. *Radiol. Case Rep.* **2016**, *11*, 450–455. [CrossRef]

21. Mariniello, G.; Napoli, M.; Russo, C.; Briganti, F.; Giamundo, A.; Maiuri, F.; Del Basso De Caro, M.L. MRI Features of Spinal Solitary Fibrous Tumors. A Report of Two Cases and Literature Review. *Neuroradiol. J.* **2012**, *25*, 610–616. [CrossRef] [PubMed]

22. Alleyne, C.H.; Cawley, C.M.; Shengelaia, G.G.; Barrow, D.L. Microsurgical Anatomy of the Artery of Adamkiewicz and Its Segmental Artery. *Neurosurg. Focus* **1998**, *5*, E2. [CrossRef]

23. Santillan, A.; Zink, W.; Lavi, E.; Boockvar, J.; Gobin, Y.P.; Patsalides, A. Endovascular Embolization of Cervical Hemangiopericytoma with Onyx-18: Case Report and Review of the Literature. *J. Neurointerv. Surg.* **2011**, *3*, 304–307. [CrossRef] [PubMed]

24. El Hindy, N.; Ringelstein, A.; Forsting, M.; Sure, U.; Mueller, O. Spinal Metastasis from Malignant Meningeal Intracranial Hemangiopericytoma: One-Staged Percutaneous OnyxTM Embolization and Resection—A Technical Innovation. *World J. Surg. Oncol.* **2013**, *11*, 152. [CrossRef] [PubMed]

25. Millesi, M.; Kiesel, B.; Woehrer, A.; Hainfellner, J.A.; Novak, K.; Martínez-Moreno, M.; Wolfsberger, S.; Knosp, E.; Widhalm, G. Analysis of 5-Aminolevulinic Acid-Induced Fluorescence in 55 Different Spinal Tumors. *Neurosurg. Focus* **2014**, *36*, E11. [CrossRef] [PubMed]

26. Apra, C.; Mokhtari, K.; Cornu, P.; Peyre, M.; Kalamarides, M. Intracranial Solitary Fibrous Tumors/Hemangiopericytomas: First Report of Malignant Progression. *J. Neurosurg.* **2018**, *128*, 1719–1724. [CrossRef]

27. Demicco, E.G.; Wagner, M.J.; Maki, R.G.; Gupta, V.; Iofin, I.; Lazar, A.J.; Wang, W.-L. Risk Assessment in Solitary Fibrous Tumors: Validation and Refinement of a Risk Stratification Model. *Mod. Pathol.* **2017**, *30*, 1433–1442. [CrossRef]

28. Apra, C.; Guillemot, D.; Frouin, E.; Bouvier, C.; Mokhtari, K.; Kalamarides, M.; Pierron, G. Molecular Description of Meningeal Solitary Fibrous Tumors/Hemangiopericytomas Compared to Meningiomas: Two Completely Separate Entities. *J. Neurooncol.* **2021**, *154*, 327–334. [CrossRef]

29. Nakada, S.; Minato, H.; Nojima, T. Clinicopathological Differences between Variants of the NAB2-STAT6 Fusion Gene in Solitary Fibrous Tumors of the Meninges and Extra-Central Nervous System. *Brain Tumor Pathol.* **2016**, *33*, 169–174. [CrossRef]

30. Reisenauer, J.S.; Mneimneh, W.; Jenkins, S.; Mansfield, A.S.; Aubry, M.C.; Fritchie, K.J.; Allen, M.S.; Blackmon, S.H.; Cassivi, S.D.; Nichols, F.C.; et al. Comparison of Risk Stratification Models to Predict Recurrence and Survival in Pleuropulmonary Solitary Fibrous Tumor. *J. Thorac. Oncol.* **2018**, *13*, 1349–1362. [CrossRef]

31. Barthelmeß, S.; Geddert, H.; Boltze, C.; Moskalev, E.A.; Bieg, M.; Sirbu, H.; Brors, B.; Wiemann, S.; Hartmann, A.; Agaimy, A.; et al. Solitary Fibrous Tumors/Hemangiopericytomas with Different Variants of the NAB2-STAT6 Gene Fusion Are Characterized by Specific Histomorphology and Distinct Clinicopathological Features. *Am. J. Pathol.* **2014**, *184*, 1209–1218. [CrossRef] [PubMed]

32. Haas, R.L.; Walraven, I.; Lecointe-Artzner, E.; van Houdt, W.J.; Scholten, A.N.; Strauss, D.; Schrage, Y.; Hayes, A.J.; Raut, C.P.; Fairweather, M.; et al. Management of Meningeal Solitary Fibrous Tumors/Hemangiopericytoma; Surgery Alone or Surgery plus Postoperative Radiotherapy? *Acta Oncol.* **2021**, *60*, 35–41. [CrossRef] [PubMed]

33. Haas, R.L.; Walraven, I.; Lecointe-Artzner, E.; van Houdt, W.J.; Strauss, D.; Schrage, Y.; Hayes, A.J.; Raut, C.P.; Fairweather, M.; Baldini, E.H.; et al. Extrameningeal Solitary Fibrous Tumors-Surgery Alone or Surgery plus Perioperative Radiotherapy: A Retrospective Study from the Global Solitary Fibrous Tumor Initiative in Collaboration with the Sarcoma Patients EuroNet. *Cancer* **2020**, *126*, 3002–3012. [CrossRef] [PubMed]

34. Dufour, H.; Métellus, P.; Fuentes, S.; Murraciole, X.; Régis, J.; Figarella-Branger, D.; Grisoli, F. Meningeal Hemangiopericytoma: A Retrospective Study of 21 Patients with Special Review of Postoperative External Radiotherapy. *Neurosurgery* **2001**, *48*, 756–762; discussion 762–763.

35. Apra, C.; Alentorn, A.; Mokhtari, K.; Kalamarides, M.; Sanson, M. Pazopanib Efficacy in Recurrent Central Nervous System Hemangiopericytomas. *J. Neurooncol.* **2018**, *139*, 369–372. [CrossRef]

36. Stacchiotti, S.; Tortoreto, M.; Baldi, G.G.; Grignani, G.; Toss, A.; Badalamenti, G.; Cominetti, D.; Morosi, C.; Dei Tos, A.P.; Festinese, F.; et al. Preclinical and Clinical Evidence of Activity of Pazopanib in Solitary Fibrous Tumour. *Eur. J. Cancer* **2014**, *50*, 3021–3028. [CrossRef]

37. de Bernardi, A.; Dufresne, A.; Mishellany, F.; Blay, J.-Y.; Ray-Coquard, I.; Brahmi, M. Novel Therapeutic Options for Solitary Fibrous Tumor: Antiangiogenic Therapy and Beyond. *Cancers* **2022**, *14*, 1064. [CrossRef]

38. Kawashima, M.; Suzuki, S.O.; Yamashima, T.; Fukui, M.; Iwaki, T. Prostaglandin D Synthase (Beta-Trace) in Meningeal Hemangiopericytoma. *Mod. Pathol.* **2001**, *14*, 197–201. [CrossRef]
39. Peyre, M.; Salaud, C.; Clermont-Taranchon, E.; Niwa-Kawakita, M.; Goutagny, S.; Mawrin, C.; Giovannini, M.; Kalamarides, M. PDGF Activation in PGDS-Positive Arachnoid Cells Induces Meningioma Formation in Mice Promoting Tumor Progression in Combination with Nf2 and Cdkn2ab Loss. *Oncotarget* **2015**, *6*, 32713–32722. [CrossRef]

Review

Novel Therapeutic Options for Solitary Fibrous Tumor: Antiangiogenic Therapy and Beyond

Axel de Bernardi [1], Armelle Dufresne [1], Florence Mishellany [2], Jean-Yves Blay [1,3], Isabelle Ray-Coquard [1,3] and Mehdi Brahmi [1,*]

[1] Department of Medical Oncology, Centre Léon Bérard, 28 Rue Laennec, 69008 Lyon, France; debernardi.axel@gmail.com (A.d.B.); armelle.dufresne@lyon.unicancer.fr (A.D.); jean-yves.blay@lyon.unicancer.fr (J.-Y.B.); isabelle.ray-coquard@lyon.unicancer.fr (I.R.-C.)

[2] Department of Pathology, Centre Jean Perrin, 58 Rue Montalembert, 63011 Clermont-Ferrand, France; florence.mishellany@clermont.unicancer.fr

[3] Faculty of Medicine, Université Claude Bernard Lyon 1, 69003 Lyon, France

* Correspondence: mehdi.brahmi@lyon.unicancer.fr; Tel.: +33-(0)4-78-78-51-26

Simple Summary: In the latest WHO classification, solitary fibrous tumors (SFTs) are now subdivided into benign SFT (intermediate category (locally aggressive)), SFT NOS (intermediate category (rarely metastasizing)), and malignant SFT. Thanks to recent progress in molecular characterization, the identification of the *NAB2–STAT6* fusion oncogene has emerged as a specific cytogenetic hallmark for SFT. Despite these recent advances in classification and understanding of the molecular pathophysiology of SFT, there are no consensus clinical guidelines regarding systemic treatment. Several new therapeutic options are of interest in this subtype of sarcoma considered as refractory to classical chemotherapy. In case of advanced disease, antiangiogenic therapy might be viewed as the best therapeutic option.

Abstract: SFT is an ultrarare mesenchymal ubiquitous tumor, with an incidence rate <1 case/million people/year. The fifth WHO classification published in April 2020 subdivided SFT into three categories: benign (locally aggressive), NOS (rarely metastasizing), and malignant. Recurrence can occur in up to 10–40% of localized SFTs, and several risk stratification models have been proposed to predict the individual risk of metastatic relapse. The Demicco model is the most widely used and is based on age at presentation, tumor size, and mitotic count. Total en bloc resection is the standard treatment of patients with a localized SFT; in case of advanced disease, the clinical efficacy of conventional chemotherapy remains poor. In this review, we discuss new insights into the biology and the treatment of patients with SFT. *NAB2–STAT6* oncogenic fusion, which is the pathognomonic hallmark of SFT, is supposedly involved in the overexpression of vascular endothelial growth factor (VEGF). These specific biological features encouraged the successful assessment of antiangiogenic drugs. Overall, antiangiogenic therapies showed a significant activity toward SFT in the advanced/metastatic setting. Nevertheless, these promising results warrant additional investigation to be validated, including randomized phase III trials and biological translational analysis, to understand and predict mechanisms of efficacy and resistance. While the therapeutic potential of immunotherapy remains elusive, the use of antiangiogenics as first-line treatment should be considered.

Keywords: SFT; rare sarcoma; antiangiogenics; immune-checkpoint inhibitors

Citation: de Bernardi, A.; Dufresne, A.; Mishellany, F.; Blay, J.-Y.; Ray-Coquard, I.; Brahmi, M. Novel Therapeutic Options for Solitary Fibrous Tumor: Antiangiogenic Therapy and Beyond. *Cancers* **2022**, *14*, 1064. https://doi.org/10.3390/cancers14041064

Academic Editor: Bahil Ghanim

Received: 30 December 2021
Accepted: 16 February 2022
Published: 20 February 2022

Publisher's Note: MDPI stays neutral with regard to jurisdictional claims in published maps and institutional affiliations.

1. Introduction

The fifth edition of the World Health Organization (WHO) classification of tumors of soft tissue and bone was published in April 2020, with more than 150 histological subtypes. The classification has been updated with the identification of new distinct molecular subtypes, thanks to the easier access to molecular tools, and these improvements have led to a better understanding of the tumorigenesis of many sarcomas. The classification of conjunctive tumors is crucial, not only for diagnosis and prognostication but also for correct management of patients.

In the fibroblastic and myofibroblastic tumor category, the recognized entity of solitary fibrous tumor (SFT) is described. SFT is a ubiquitous fibroblastic mesenchymal neoplasm originally reported by Klemperer et al. in 1931 [1], with locally invasive properties and a predilection for body cavity sites and serosal membranes; it has only recently been reported in the pleura and/or lung. SFT is known as the "great simulator" of soft-tissue neoplasm due to its many differential diagnoses [2]. The diagnosis of SFT is established by the conjunction of clinical, pathological, immunohistochemical, and molecular features. Identification of the *NAB2* (NGFI-A-binding protein 2)–*STAT6* (signal transduction and activator of transcription 6) fusion oncogene has emerged as a specific cytogenetic hallmark for SFT. Several risk stratification systems have been proposed for SFT to predict which tumors may harbor an aggressive behavior [2–5]. In the latest WHO classification, SFTs are now subdivided into benign SFT (intermediate category (locally aggressive)), SFT NOS (intermediate category (rarely metastasizing)), and malignant SFT [6].

In this review, we discuss recent progresses in the molecular characterization and therapeutics of SFT and future prospects of tailored therapeutic approaches.

2. Description of SFT

2.1. Epidemiology

SFT is an ultrarare mesenchymal tumor that represents 3.7% of all soft-tissue sarcomas (STSs) and mesenchymal tumors [7] with a reported incidence rate <1 case/million people/year [8,9]. SFT usually affects adult patients with a peak incidence in the fifth and sixth decades. Pleural SFTs are observed in older patients (median age at diagnosis 56 to 60) compared with intra-abdominal or meningeal SFTs that occur in younger patients (fourth decade) [10].

2.2. Clinical and Radiological Presentation

The thorax (pleura, mediastinum, lung parenchyma) is the most common site of initial presentation (30%), but SFT can occur at any extra-thoracic anatomical site [11]. The most frequent extra-pleural location is represented by the head and neck, including meninges (27%), followed by the intra-abdominal region (peritoneum, retroperitoneum, and pelvis) (20%), trunk (10%), and extremities (8%) [12]. SFTs localized in the retroperitoneum, peritoneum, or mediastinum tend to have a more aggressive course compared to other anatomical sites [5,13,14]. SFTs are often slow-growing tumors, with delayed symptom onset. Patients may present with nonspecific pulmonary symptoms such as dyspnea or cough, but often remain asymptomatic despite important tumor volume. Compression of adjacent anatomical structures can be observed for voluminous masses but remains unusual as the median tumor size at diagnosis ranges between 50 and 80 mm [15]. Additionally, few patients (<10%) develop paraneoplastic syndromes which can guide the diagnosis. Hypertrophic osteoarthropathy (Pierre Marie–Bamberger syndrome) is a nonspecific condition rarely associated with pleural SFT. Classical symptoms include distal digital clubbing, periostitis, and synovial effusions, supposedly linked to the paraneoplastic overexpression of vascular endothelial growth factor (VEGF). Fewer than 5% of SFT patients can also present with a refractory hypoglycemic syndrome due to the overproduction of insulin-like growth factor-2 (IGF-2) by large peritoneal or pleural SFTs, called Doege–Potter syndrome [16–18].

Consequently, SFTs are frequently diagnosed on incidental radiological findings [19]. Time to diagnosis is classically longer in pleural SFTs compared to extra-pleural SFTs. The radiographic features of SFTs are nonspecific, and computed tomography (CT) usually shows a well-circumscribed isodense mass to skeletal muscle with contrast enhancement in highly vascularized tumors (65%) [15]. SFTs are often characterized by the presence of low-signal-intensity foci on T1- and T2-weighted magnetic resonance imaging (MRI), corresponding to the collagen content. ^{18}F-FDG PET/CT may have a limited role in diagnosing SFT in suspected patients, although the presence of multiple high-grade lesions associated with ^{18}F-FDG PET hypermetabolism should raise suspicion of a malignant SFT.

2.3. Biopathology

2.3.1. Classification

SFTs belong to the group of fibroblastic and myofibroblastic tumors in the WHO classification [6]. The fifth edition published in April 2020 subdivided SFTs into three categories benign (locally aggressive), NOS (rarely metastasizing), and malignant [20]. Meningeal SFT, previously known as hemangiopericytoma (HPC), is a rare form of extra-pleural SFT that derives from smooth muscle pericytes surrounding the intraparenchymal microvasculature [21]. Historically, SFTs were considered low-grade tumors contrary to HPCs that displayed aggressive patterns. Despite an apparent distinct clinical behavior, SFTs and HPCs were merged under one disease umbrella in the fourth WHO classification [22]. The fourth revised edition of the WHO Classification of Tumors of CNS published in 2016 introduced the hybrid "SFT/HPC" class, subdivided into three histological grades depending on mitotic count (grade 3 defined by at least five mitoses/10 HPFs) [23]. Most recently, the term of "hemangiopericytoma" was removed from the 2021 WHO Classification of Tumors of CNS to conform with soft-tissue pathology nomenclature [24]. In other words, updated classifications now consider SFTs and HPCs as the same entity at two opposite ends of the same histologic spectrum rather than the strict "benign or malignant" dichotomy that was used for decades (Figure 1).

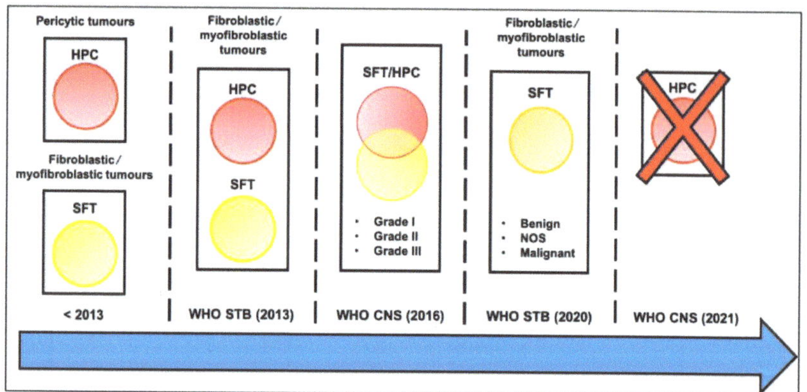

Figure 1. Evolution of the WHO classification of SFTs over time. A crucial update for SFT classification is the development of risk stratification models that resulted in improved prognostication over the traditional benign/malignant distinction (please see the section below).

2.3.2. Prognosis and Risk Stratification

Localized SFTs offer good prognosis after complete surgery. However, recurrence can occur in up to 10–25% of SFTs by 10 years [5,25–27]. The risk of metastatic recurrence at 5 years can even rise up to 40% in high-risk patients [13,28–32]. Preferential metastatic sites include lungs, liver, and bone.

SFT recurrence is more frequent in the case of incomplete resection (R1/R2) [25,28], tumor seeding within serosal membranes (pleura, peritoneum), or meningeal or distant hematogenous spread. Small retrospective series tend to show that patients with extra-pleural SFTs have a higher risk of local and metastatic recurrence compared to pleural SFTs [33–35]. Meningeal SFTs have a dismal prognosis with frequent local recurrence due to meningeal seeding and early bone metastatic recurrence. Notably, late relapse beyond 10 years and up to 20 years after initial presentation is common, which justifies a long-term follow-up [13,28–32].

Importantly, the development of multivariate risk stratification models has resulted in improved prognostication, such that the traditional benign/malignant distinction is now avoided. Among those models that integrate several clinicopathologic variables to predict the individual risk of metastatic recurrence, the Demicco model ("D-score" or "MDACC score") is the most widely used in clinical practice and is applicable to SFTs of all extra-meningeal sites [36]. It is based on age at presentation, tumor size, and mitotic count to classify SFTs with a low, moderate, or high risk of developing a metastatic recurrence [37] (Table 1). Another version also includes tumor necrosis. Importantly, dedifferentiation, which corresponds to the abrupt transition from a conventional low-grade SFT component into a high-grade sarcoma and occurs in <1% of primary or recurrent SFTs, remains unpredictable by this model [38].

Table 1. Risk stratification model proposed by Demicco et al. [2].

Risk Factor	Cut-Off	Points Assigned	
		3-Variable Model	4-Variable Model
Patient age (years)	<55	0	0
	>55	1	1
Mitoses/mm^2	0	0	0
	0.5–1.5	1	1
	≥2	2	2
Tumor size (cm)	0–4.9	0	0
	5–9.9	1	1
	10–14.9	2	2
	≥15	3	3
Tumor necrosis	<10%	N/A	0
	≥10%	N/A	1
Risk	Low	0–2 points	0–3 points
	Intermediate	3–4 points	4–5 points
	High	5–6 points	6–7 points

Ghanim et al. used blood-derived biomarkers to predict event-free-survival (EFS) in intrathoracic SFTs [39]. Elevated preoperative fibrinogen was an independent prognostic marker of poor outcome after curative surgery and associated with malignant SFT. This finding suggests a potential role of innate immune system overactivation in SFT prognosis.

The presence of *TP53* and/or *TERT* mutations seems to be correlated with malignant SFTs associated with an aggressive biological behavior.

2.3.3. Tumorigenesis, Pathology, and Molecular Alterations

The macroscopic appearance of SFT is a well-defined, lobulated, firm mass surrounded by a serosal pseudo-capsule. The wide histological spectrum of SFT ranges from morphologically paucicellular to highly cellular tumors. SFTs are composed of atypical spindle cells with fusiform nuclei arranged haphazardly with a suggestive "patternless pattern", surrounded by a dense stromal collagen with thin collagen bands (Figure 2A) and admixed with a characteristic branching staghorn (hemangiopericytoma-like) vasculature. Hypocellular phenotypes have a low mitotic count, and nuclear pleomorphism or necrosis is classically absent. These histopathologic patterns are not specific to SFTs and can also be observed in other mesenchymal tumors [40,41].

In practice, the detection of the *NAB2–STAT6* fusion gene (detailed below) with cytogenetic methods such as fluorescent in situ hybridization (FISH) is impossible due to the small size of the inverted sequence and the proximity of the *NAB2* and *STAT6* loci. For routine diagnosis, STAT6 is a robust immunohistochemical surrogate marker of all *NAB2–STAT6* fusion transcripts. A strong nuclear expression of the C-terminal part of STAT6 has good diagnostic performance with excellent sensitivity (98%) and specificity (>85%) [42–46] (Figure 2B). PCR-based detection of STAT6 is less sensitive for diagnosis due to the diversity of possible breakpoints in fusion transcripts. However, STAT6 is inconsistently expressed and might be absent in some SFT cases. STAT6 is rarely overexpressed in the nucleus and cytoplasm of well-differentiated (WD-LS) and dedifferentiated liposarcoma (DD-LPS) cells [47]. If that differential diagnosis is discussed, a complementary analysis of MDM2 and CDK4 expression in IHC can be useful as they are overexpressed in WD-LS/DD-LPS but not in SFT. Other standard IHC markers can be used in combination (CD34, Bcl2, CD99, vimentin, desmin, S100 protein, epithelial markers) with good sensitivity but a low specificity. Lastly, GRIA2 and ALDH1 are under investigation [48,49].

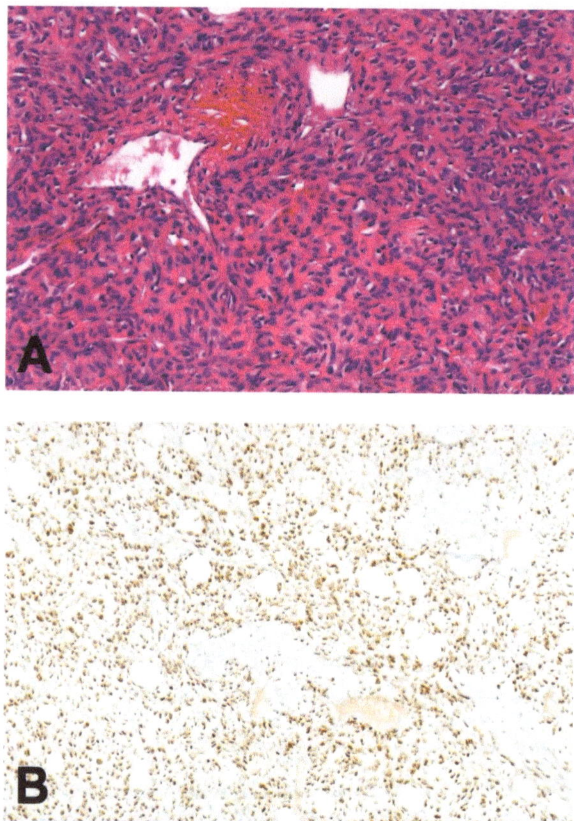

Figure 2. (**A**) Morphological appearance of SFT composed of spindle cells with a patternless architecture and a dense hyalinized collagenous stroma. Magnification: ×40; staining: hematoxylin, eosin, and Safran (HES). (**B**) Diffuse and strong STAT6 nuclear staining in SFT tumor cells. Magnification: ×40; staining: STAT6 antibody.

The WHO classification updates were justified by the discovery of a shared cytogenetic signature across all anatomical sites in 2013 by three different research groups: the *NAB2–STAT6* fusion oncogene [41,50–52]. The *NAB2–STAT6* fusion gene is detected in nearly all SFT cases, which suggests a driver role in tumor development. *NAB2* and *STAT6* are adjacent genes transcribed in opposite directions at the locus 12q13 [41,51,52].

NAB2 is a transcriptional repressor that binds to the inhibitory domain of *EGR-1* (early growth response 1) through its N-terminal binding domain (NAB2 conserved domain 1—NCD1) to have transcriptional control over EGR-1 target genes, including IGF-2 or FGFR1. *EGR-1* is a downstream gene of the MAPK/ERK pathway [53]. EGR-1 enhances the transcription of cell-cycle regulatory proteins such as cyclin D1 that promotes tumor cell proliferation, which in turns activates the MAPK/ERK signaling pathway and auto-activates the expression of EGR-1 through a positive feedback loop. Additionally, the transcription of *EGR-1* is upregulated by various growth factors such as IGF-1 and its receptor IGFR-1. *EGR-1* also plays a role in the systemic dissemination of tumor cells. Indeed, epithelial–mesenchymal transition (EMT) was reported to be triggered by the *EGR-1*-induced upregulation of Slug and Snail via the ERK1/2 and PI3K/Akt pathways in ovarian cancer cells [54]. Proangiogenic growth factors such as basic fibroblast growth factor (bFGF) and VEGF-A are EGR-1 target genes that contribute to tumor angiogenesis. EGR-1 can directly activate the transcription of these factors [55] or be stimulated via the ERK1/2 pathway after hypoxia-inducible factor (HIF)-1α overexpression in response to hypoxia in the tumoral microenvironment.

STAT6 (signal transducer and activator of transcription 6) is a transcription factor usually involved in allergic and immune signaling pathways with a role in tumorigenesis [56,57]. The Src homology 2 domain (SH2) is crucial to bind to IL-4R or IL-13R and trigger STAT6 phosphorylation by Janus (JAK) and TyK2 kinases [58]. In its dimerized form, STAT6 is activated, whereby it can enter the nucleus and bind to DNA promoters through its DNA-binding domain (DBD1).

In SFT, the recurrent intrachromosomal inversion in the long arm of chromosome 12 leads to the replacement of at least one of the three repressor domains of *NAB2* (NCD1, NCD2, CID) by the transactivation domain (TAD) of *STAT6*. *NAB2* and *STAT6* are fused in a common direction of transcription, which results in the transcription of a chimeric NAB2–STAT6 fusion protein from the *NAB2* promoter [50,52]. Then, the fusion transcript translocates to the nucleus and constitutively activates EGR-1-responsive genes [41] (Figure 3). Intriguingly, *NAB2* and *EGR-1* are mutual targets to each other, which creates a positive feedback loop and strengthens the abnormal accumulation of the *NAB2–STAT6* fusion transcript in the nucleus of SFT cells. This specificity is a hallmark of SFT used for differential diagnosis with other tumors.

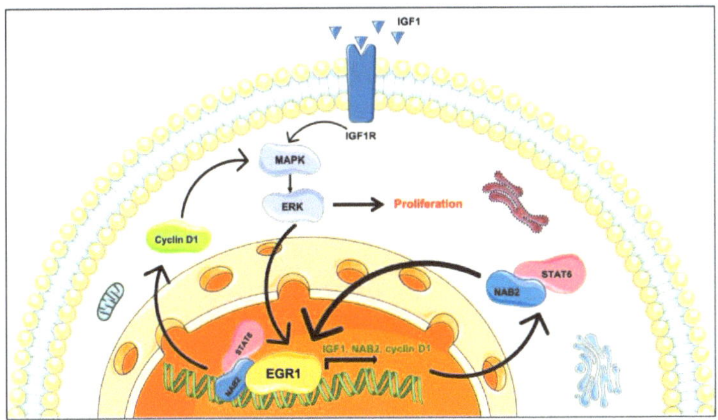

Figure 3. Suspected role of the NAB2–STAT6 fusion transcript in SFT tumorigenesis.

Barthelmess and colleagues first suggested that the *NAB2 exon4–STAT6 exon2* (N4S2) and the *NAB2 exon6–STAT6 exon16/17* (N6S16/17) fusion variants might be associated with distinct clinical features. *N4S2* was mostly found in older patients with less aggressive SFTs and deep extra-thoracic lesions. On the other hand, *N6S16/17* was more frequent in younger patients with aggressive phenotypes and clinical behavior usually found in meningeal SFTs [50]. Since then, other studies failed to demonstrate a clear impact of fusion variants on prognosis, probably due to a short follow-up that did not take into account late recurrences [59–63]. In a recent retrospective cohort with long-term follow-up, Georgiesh et al. investigated the clinicopathological and prognostic impact of the STAT6-Full (intact STAT6 domains) and STAT6-TAD (contains only the STAT6 TAD domain) variants by

RNA sequencing [64]. A total of 39 patients with localized extra-meningeal SFTs were enrolled. Patients with STAT6-TAD tumors had a worse prognosis, with a higher mitotic count and a 10 year recurrence-free survival rate of 25% (vs. 78% for STAT6-Full patients). These promising results need further confirmation in prospective trials to conclude on their prognostic value.

3. Therapeutic Options for SFT

Patients with SFTs should be managed within sarcoma reference centers, by a dedicated multidisciplinary team with a pathologist, radiologist, surgical oncologist, radiation oncologist, and medical oncologist who are familiar with the nuances of this disease [65–68]. Each case has to be discussed in a specialized multidisciplinary tumor board (MTB) to determine the best individualized therapeutic strategy.

For response assessment, the use of both Response Evaluation Criteria in Solid Tumors (RECIST) and the new Choi criteria (i.e., a 10% decrease in tumor size or a more than 15% decrease in tumor density) might be of interest to evaluate therapeutic response to antiangiogenic agents [69–71]. Choi criteria were originally developed to predict the response of advanced gastrointestinal stromal tumor (GIST) to imatinib. Due to the hypervascularized nature of SFT, recent studies used both Choi and RECIST criteria to assess radiological response. The retrospective studies detailed below reported few responses with RECIST criteria but 46% to 79% response with Choi criteria [72–76]. However, the predictive value of Choi criteria in SFTs should be interpreted cautiously. There are no data on the use of Choi criteria in patients with advanced SFTs treated with chemotherapy.

3.1. Localized Disease

3.1.1. Surgery

Complete en bloc surgical resection with negative margins (R0) is the gold-standard treatment for localized disease [77]. The 10 year overall survival (OS) in patients with SFTs resected with negative margins ranges from 54% to 89% [78–80]. In pleural SFTs, the modalities of resection include wedge resection, lobectomy, or pneumonectomy, associated with chest wall or diaphragm resection if necessary [77,81,82]. In their retrospective series, Lahon et al. resected localized pleural SFTs in 157 patients. Despite R0 margins, 15 patients (10%) recurred with a median time to recurrence of 29 months (10 local and five metastatic recurrences) [81]. Re-resection of local recurrence could achieve local control in 9/10 patients. The 5 year and 10 year OS rates were 86% and 77%, respectively. Similarly, Lococo et al. reported disease recurrence in 15/50 patients (six local and nine metastatic recurrences) with localized pleural SFTs resected with negative margins, yielding 5 year and 10 year OS rates of 81% and 67%, respectively [83]. Complete surgical resection is also the primary approach in extra-pleural tumors, with procedures paralleling other sarcoma surgeries from similar anatomical sites.

3.1.2. Radiation Therapy

Several strategies can be discussed in the postoperative setting, depending on surgical margins and risk stratification models. Despite limited available prospective data, multiple observational studies report improved local control in high-risk SFT patients treated with adjuvant radiotherapy (RT). Nevertheless, there is no clear demonstration of an OS benefit of adjuvant RT [13,14,26,33,34,84–88]. In 2020, Haas et al. investigated the role of perioperative RT in localized extra-meningeal SFT [14]. In this cohort, 428/549 patients (78%) received surgery alone and 121 (22%) were treated with surgery and RT. Overall, 48% of patients (n = 58) received surgery followed by adjuvant RT, with a lower risk of local progression (p = 0.012), and a 96% local control rate at 5 years. However, this association did not translate into an OS benefit (p = 0.325). Preoperative RT was assessed in the STRASS phase III randomized trial, which included intra-abdominal sarcomas of different histotypes, but SFT cases were underrepresented, leading to inconclusive results in that population [88,89]. An international retrospective observational study performed across seven specialized sarcoma centers aimed to better define the benefit of definitive RT in this disease [90]. Forty patients with locally advanced or metastatic SFT (nine pleural, 16 soft-tissue, 10 meningeal, three head and neck, and two other SFTs) were treated with definitive RT, receiving approximately 60 Gy, with an objective response rate of 67%. At 5 years, the local control (LC) rate was 81.3%, and the OS rate was 87.5%. In case of palliative RT (39 Gy in conventional fractionation), the overall response rate (ORR) was 38%, and the 5 year LC and OS rates were 62.5% and 54.2% respectively. Therefore, radiation doses can range from 39 Gy (in conventional fractions) for palliative care to 60 Gy to achieve durable control.

In summary, observation is recommended in patients with negative margins (R0) without high-risk histologic features. In the case of intermediate- to high-risk SFT with positive margins (R1/R2), re-resection should be discussed for fit patients if complete resection can be achieved with minimal morbidity. If the patient is unfit for further resection or R0 surgery cannot be technically achieved (due to anatomical site), then adjuvant RT is a reasonable option. Surgeons may adopt a conservative attitude in selected cases to preserve organ functions when postoperative RT can be delivered, given the favorable long-term outcome. Adjuvant RT should systematically be considered for high-risk SFTs, such as malignant SFTs of the central nervous system (CNS), considering the very high risk of local recurrence. Neoadjuvant RT can be an option in selected cases to improve tumor resectability or when wound complications are predicted to be manageable.

3.1.3. (Neo)Adjuvant Chemotherapy

In patients with localized, resectable SFTs, there is no evidence supporting the use of systemic therapies in the (neo)adjuvant setting [18]. A review of the literature shows only limited data on this subject and consists mostly of a few case reports [91–93]. Even if the relevance of adjuvant chemotherapy (CT) in resected SFTs is still unknown due to a lack of data, it is important to note that SFTs marginally benefit from traditional sarcoma CT with low response rates (Table 2). Importantly, adjuvant CT should never be intended to rescue inadequate surgery. Nevertheless, for some patients with high-risk SFTs and/or large malignant tumors, the use of neoadjuvant CT should be discussed at a specialized MTB. Importantly, eligible patients should be managed within clinical studies. Therefore, in the case of locally advanced tumors, if R0 surgery is not feasible apart from mutilating surgery, neoadjuvant CT is an option. On the basis of the response rate data (Table 2), an anthracycline-based regimen (plus ifosfamide or dacarbazine) can be considered as the regimen of choice. Importantly, early tumor response assessment is required to avoid delaying surgery in the case of nonresponding disease.

3.2. Advanced and Metastatic Disease

3.2.1. Surgery, Ablations, or RT

Metachronous (disease-free interval >1 year), resectable lung metastases without extrapulmonary disease may be managed with surgery as standard treatment applied to sarcoma, if complete excision of all lesions is feasible. Surgery, ablations, or RT of extrapulmonary metastases may also be an option in highly selected cases [65–67].

3.2.2. Chemotherapy

In cases of synchronous and/or unresectable lung metastases and in cases of extrapulmonary metastatic disease, patients are candidates for systemic treatment, even though a standard medical approach is currently not established. Published data on the response of SFTs to conventional chemotherapy are limited, and results gathering small case series, retrospective studies, and predictive preclinical models show conflicting results. Table 2 summarizes the data available on the efficacy of systemic agents in SFT.

Table 2. Summary of available data on systemic agents in SFT.

Authors	Design	n	Drug (n)	Response/Duration/RR	mPFS (Months)	mOS (Months)
Constantinidou et al. 2012 [94]	R	24	Anthracycline-based (75 mg/m^2) (n = 17) Non-anthracycline-based (n = 7)	– 1 PR, 40% SD, 50% PD – 5 SD, 2 PD	4.2 (95% CI: 0–10.1)	14.6 (95% CI: 9.3–19.9)
Levard et al. 2013 [95]	R	23	Doxorubicin alone (n = 9) Pegylated liposomal doxorubicin (n = 1) Doxorubicin + ifosfamide (n = 8) Doxorubicin + palifosfamide (n = 1). Vinorelbine (n = 1) Paclitaxel (n = 1) Carboplatin and paclitaxel (n = 1) Brostallicin (n = 1)	2 PR (8.7%) in doxorubicin-based, 13 SD (59%), 8 PD (35%). 9 pts (39%) progression-free at 6 months	5.2 (95% CI: 3.2–7.1)	33.5 (95% CI: 14.2–52.8) (whole cohort, n = 30)
Stacchiotti et al. 2013 [74]	R	31	Anthracycline monotherapy (n = 8), Anthracycline + ifosfamide (n = 23) Ifosfamide (n = 19)	– RECIST: 6 PR (20%), 8 SD (27%), 16 PD (53%). 20% progression-free at 6 months. – 2 PR (10%), 5 SD (26%), 12 PD (63%)	4 (range 2–15)	11.5 (range 3–50)
Stacchiotti et al. 2013 [96]	R	8	Dacarbazine monotherapy (n = 8), (1200 mg/m^2)	RECIST: 3 PR, 4SD, 1PD	7 (range 2–12)	-
Park et al. 2013 [72]	R	21	Doxorubicin-based (n = 15) Gemcitabine-based (n = 5) Paclitaxel (n = 5)	0 PR, 16 SD (89%), 2 PD (11%). 5 (28%) progression-free >6 months	4.6 (95% CI: 4.0–5.3)	10.3 years (95% CI: 5.7–14.9 years)
Khalifa et al. 2015 [97]	R	11	Trabectedin (1.5 mg/m^2)	RECIST: 1 PR (9.1%), 8 SD (72.7%)	11.6 (95 % CI = 2.0; 15.2)	22.3 (95 % CI = 9.1; NR)
Le Cesne et al. 2015 [98]	R	13	Trabectedin 1.5 mg/m^2	-	7.6 (95% CI: 1.6–13.7)	14.3 (95% CI: 0.8–27.8)
Schöffski et al. 2020 [99]	R	26	Doxorubicin (n = 15) (57.7%) Doxorubicin + ifosfamide (n = 3) (11.5%) Doxorubicin/olaratumab (n = 2) (7.7%) Doxorubicin + evofosfamide (n = 1) Doxorubicin + ifosfamide + cisplatin (n = 1) *2nd line*: ifosfamide (n = 5), pazopanib (n = 5)	Doxorubicin-based: 2 PR (13.3%), 4 SD (26.7%) Ifosfamide: no response Pazopanib: 3 PR (60%)	34.1 (95% CI: 1.0–157.1)	56.0 (95% CI: 0.3–258.3)
Outani et al. 2020 [100]	R	31	Anthracycline-based (n = 11) Eribulin mesylate (n = 4) Gemcitabine + docetaxel (n = 10) Ifosfamide-based (n = 7) Pazopanib (n = 22) Trabectedin (n = 6) Other (n = 13) Multidrug regimen (n = 21)	-	-	55 (95% CI: 40–86)
Kobayashi et al. 2020	R	140	Trabectedin (1.2 mg/m^2)	11 PR (7.9%), 54 SD (41.9%) with 25/54 SD >6 months	3.7 (95% CI: 2.8–5.7)	16.4 (95% CI: 11.5–21.2)
Mulamalla et al. 2008 [101]	CR	1	Sunitinib	SD for 6 months	-	-
De Pas et al. 2008 [102]	CR	1	Imatinib	PR for 21 months with major clinical benefit	-	-
George et al. 2009 [103]	R	48	Sunitinib (37.5 mg)	1 PR, 11 SD + PR (22%) at 16 weeks, 7 (14%) at 24 weeks.	1.8	-
Domont et al. 2009 [104]	P	2	Dacarbazine (1000 mg/m^2) + sorafenib (400 mg) Sunitinib (50 mg)	PR, PD at 1.5 year SD > 6 months	-	-
Park et al. 2011 [73]	R	14	Temozolomide (150 mg/m^2)–bevacizumab (5 mg/kg)	Choi: 11 PR (79%), 2 SD	9.7 (95% CI: 7.31–not estimable)	-

Table 2. *Cont.*

Authors	Design	n	Drug (n)	Response/Duration/RR	mPFS (Months)	mOS (Months)
Valentin et al. 2013 [105]	P	5	Sorafenib (800 mg)	No objective response, 2/5 SD (9 months)	-	19.7
Levard et al. 2013 [95]	R	10	Pazopanib (800 mg) (n = 6) Sunitinib (37.5 mg) (n = 4)	No objective response, 5 SD, 5 PD 4 pts (40%) progression-free >6 months: at 8.0 and 14.0 (pazopanib), and 29.5 and 29.9 months (sunitinib).	5.1 (95% CI: 0.7–9.6).	-
Stacchiotti et al. 2010 [106]	R	11	Sunitinib (37.5 mg)	6 PR, 1 SD (Choi)Response >6 months in 5 pts	-	-
Stacchiotti et al. 2014 [107]	R	6	Pazopanib (800 mg)	No RECIST response 1 PR, 2 SD (Choi)	3 (range 1–15)	-
Maruzzo et al. 2015 [76]	P	13	Pazopanib 1st line (800 mg)	5 PR (46%), 4 SD (36%) (Choi)	4.7 (95% CI: 4.8–7.4)	13.3 (95% CI: 3.9–22.6)
Martin-Broto et al. 2019 [108]	P	36	Pazopanib (800 mg)	Choi: 18 PR (51%), 9 SD (26%) RECIST: 2 PR (6%), 21 SD (60%)	5.6 (95% CI: 4.51–6.62)	NR
Martin-Broto et al. 2020 [109]	P	31	Pazopanib (800 mg)	18 PR (58%), 12 SD (39%), 1 PD (3%)	12.1 (range 2.6–21.7)	49.8 (range 8.2–91.3)

(R = retrospective study; P = prospective study; CR = case report; pts = patients; CI = confidence interval; NR = not reached).

For anthracycline-based therapies, ORR ranges between 0% and 20%, and stable disease (SD) is achieved in 26–65% of cases with a median PFS (mPFS) and median OS (mOS) of 4–5.2 months and 11.5–14.6 months, respectively [72,74,94,95,99]. Ifosfamide monotherapy as first-line treatment was also assessed with a 10% ORR and 26% SD. Dacarbazine monotherapy showed a 37.5% ORR and an mPFS of 7 months [96]. Despite a 7.9–9.1% ORR, trabectedin is an important alternative in subsequent lines, as disease stabilization can be achieved in 42–73% of cases with an mPFS of 3.7–11.6 months and an mOS of 14.3–22.3 months [97,98,110]. Consequently, the use of trabectedin is often associated with a meaningful clinical benefit in patients with advanced SFTs. Subsequent lines of chemotherapy consisted of cytotoxic agents (in monotherapy or in combination) used routinely for STS treatment, but showed no objective response.

Common and frequent chemotherapy-induced adverse events such as high-grade mucositis, nausea and vomiting, or hematologic toxicities were reported with all the regimens. Additionally, patients could experience congestive heart failure (associated or not with a reduction in cardiac ejection fraction) with anthracycline, ifosfamide-induced encephalopathy or hepatotoxicity, and rhabdomyolysis with trabectedin. The toxicity profiles were similar to those classically observed in STS.

Overall, the consistent response rates observed across retrospective series tend to confirm the limited efficacy of conventional chemotherapy in advanced SFT. The STRADA randomized phase II study (NCT03023124) will prospectively compare trabectedin (1.3–1.5 mg/m^2) to doxorubicin (75 mg/m^2) plus dacarbazine (400 mg/m^2/day, days 1, 2) in 50 patients with advanced SFTs. The Italian phase II single-arm ERASING trial (NCT03840772) will also evaluate the activity of eribulin in that population. Interestingly, the Choi response rate is one of the secondary outcomes of the ERASING and STRADA trials, which will provide data in these populations.

In summary, even if there is no formal gold standard, patients with metastatic SFTs are treated routinely in the same way as other STSs, using anthracycline as first-line therapy, while ifosfamide, dacarbazine, and trabectedin are options for second line and beyond. Best supportive care alone is an alternative for unfit patients, regarding the toxicities of those drugs. Otherwise, there is no demonstration that multiagent CT improves patient survival, and single-agent chemotherapy remains the standard.

3.2.3. Antiangiogenic Therapies and Other Targeted Therapies

SFTs are highly vascularized tumors with high expression rates of proteins involved in angiogenic pathways such as platelet-derived growth factor receptor (PDGFR) and vascular endothelial growth factor receptor (VEGFR) [111]. Angiogenesis is an important feature involved in tumor growth and metastatic diffusion of SFT. Accordingly, the inhibition of angiogenesis pathways was suspected to be a key therapeutic target to inhibit tumor cell proliferation. Considering the angiogenic properties of SFTs, several case reports and clinical trials investigated various inhibitors of angiogenesis with a promising activity in SFTs.

Prolonged disease control up to 30 months has been reported with antiangiogenic agents (sunitinib, sorafenib, pazopanib, temozolomide–bevacizumab) in case reports, retrospective studies, and phase I and phase II trials [76,95,101–106]. ORR varied from 0% to 79% with an mPFS ranging between 4.7 months and 9.7 months using Choi criteria. Antiangiogenic agents were positioned in subsequent lines after chemotherapy resistance.

A prospective single-arm phase II trial confirmed the efficacy of pazopanib in two cohorts of typical ($n = 34$) and malignant/dedifferentiated (DD) SFTs ($n = 36$) with 58% and 51% partial response (PR), respectively, according to Choi criteria. [108,109]. Interestingly, pazopanib was associated with a very high clinical benefit rate in the typical SFT cohort (97%). The outcomes of PFS were also more favorable in the typical SFT cohort than in the malignant/DD SFT cohort, with an mPFS of 9.8 month using Choi criteria and 12.1 months based on RECIST criteria in the typical SFT cohort, versus 5.6 months for the malignant/DD SFT group. Notably the inclusion of patients with DD-SFT was stopped due to cases of hyper-progression under pazopanib in a planned interim analysis. The toxicity profiles were similar between the two cohorts and consistent with those reported in previous clinical trials.

Importantly, pazopanib harbors a multifaceted cardiovascular toxicity profile including cardiomyopathy, QTc-interval prolongation, and hypertension. Therefore, a cardiologic workup should always be proposed in patients with a preexisting cardiovascular condition. Even in the absence of comorbidity, it should be considered in patients previously treated with anthracycline-based CT.

In summary, according to the aforementioned results, pazopanib can be a treatment option in the first-line setting for typical SFT.

3.2.4. IGF-1 Inhibitors

IGF-1 is overexpressed in SFT, and treatment regimens using figitumumab, a fully human IgG2 anti-IGF-1 (IGF-1R) monoclonal antibody, demonstrated tumor responses in a few patients with advanced SFTs [106,112]. In a phase I trial assessing the combination of figitumumab and the mTOR inhibitor everolimus, the only PR (among 18 evaluable patients) was observed in a patient with SFT [112]. Unfortunately, Pfizer ceased the development of the drug in 2011 and has stopped its manufacture.

3.2.5. Immunotherapy

Immunotherapy is another promising approach for SFT. Available data on the SFT immune microenvironment mainly come from retrospective studies.

Tazzari et al. first reported an immunosuppressed environment at the tumor site marked by the absence of a granulocytic MDSC (gMDSC) infiltrate [113]. Samples treated with sunitinib malate were enriched in activated CD8$^+$ and CD4$^+$ cells, which suggests that antiangiogenic therapies might modulate the T-cell immune infiltrate in SFTs. A translational study on 16 intracranial SFT/HPC specimens (13 HPCs, three SFTs) evaluated the correlation of PD-1, PD-L1, and tumor-infiltrating lymphocyte (TIL) expression with prognosis [114]. PD-L1 was expressed in all tumors. As an individual biomarker, a diffuse or intense PD-L1 staining was associated with a shorter time to treatment failure (TTF). Diffuse PD-L1 staining coupled with the absence of CD8 expression was significantly associated with a shorter TTF ($p = 0.005$). Consequently, the conjunction of diffuse PD-L1 IHC staining and the absence of CD8 expression may predict the early occurrence of extracranial metastases in SFTs [114]. Dancsok et al. systematically evaluated the expression of immune checkpoint biomarkers and TILs in a variety of sarcoma subtypes [115]. Among the 16 SFT cases included in the study, PD-1 and PD-L1 were infrequently expressed with sparse TILs. On the basis of the results in the entire cohort, the authors suggested that T-cell immune infiltrate might be less frequent in translocation-associated sarcomas, such as SFTs. More recently, Berghoff et al. investigated the inflammatory tumor microenvironment in 74 specimens of meningeal tumors including 12 cases (16.2%) of HPC and seven cases (9.5%) of meningeal SFT [116]. TILs were present in all SFT cases and 11/12 cases (91.7%) of HPC. PD-L1 was only expressed in 1/12 (8.3%) cases of HPC.

Several cases of partial response to anti-PD1 or -PDL1 therapies have been reported in the recent literature. Boothe et al. published the case of a patient with advanced malignant pleural SFT who experienced a long-lasting near-complete response after 31 cycles of pembrolizumab. PD-L1 was positive in IHC (5%), and a mutation in exon 9 of *MLH1* (E234Q) was detected after next-generation sequencing (NGS) [117]. Lastly, in a phase II trial of pembrolizumab in sarcomas, one case of SFT was the only PR reported in the study [118]. Given the suspected role of the immune system in these neoplasms, a phase III trial is currently ongoing to compare nivolumab + ipililumab to pazopanib in adults with advanced rare STSs including SFTs (NCT04741438).

4. Conclusions

Sarcomas represent a highly heterogeneous group of tumors, both in clinical and in genomic settings; thus, they should be treated separately. Furthermore, recent clinical trials exploring unselected sarcoma histotype populations [119] failed to improve patient outcome. Therefore, each subtype should still be treated separately.

SFTs are poorly sensitive to conventional chemotherapy. Nevertheless, the pathognomonic *NAB2–STAT6* oncogenic fusion that induces IGF-1 overexpression and angiogenesis in the tumor microenvironment might help considering SFT as a targetable sarcoma. Those biological insights recently translated into clinical management, especially IGF-1 inhibitor (figitumumab) and antiangiogenic drugs (including pazopanib, sunitinib and sorafenib). The results presented in this review suggest that antiangiogenic therapies such as pazopanib could be of interest for first-line treatment, while data on the efficacy of immunotherapy remain scarce, and more results are needed.

Future approaches in advanced SFT treatments should focus on international collaborations to develop large, randomized phase III trials to assess the efficacy of antiangiogenics and/or immune checkpoint inhibitors (ICI) (PD-1 inhibitors and PD-L1 inhibitors) compared to conventional chemotherapy. Additionally, a randomized phase III trial comparing antiangiogenic therapy alone versus antiangiogenic therapy plus anti-PD-1/PD-L1, as well as ICI in combination, would be of interest. For localized disease, could (neo)adjuvant therapy based on antiangiogenics and/or ICI decrease recurrence rate in high-risk patients? Nevertheless, an international approach to research on this rare disease is unavoidable, and additional biological ancillary studies are highly recommended.

In conclusion, SFT is a rare STS subtype for which standard chemotherapy has been reported to have limited efficacy. Overall, our review underlines the modest activity of standard chemotherapy in SFT but confirms that antiangiogenic agents have interesting activity and might be considered as the best therapeutic option in the advanced setting. However, the prognosis remains poor, and the inclusion of patients in clinical trials is highly recommended.

Author Contributions: Conceptualization, A.d.B., I.R.-C. and M.B.; methodology, M.B.; software, A.d.B.; validation, A.D., J.-Y.B., I.R.-C. and M.B.; formal analysis, A.d.B. and M.B.; investigation, A.d.B. and M.B.; resources, F.M.; data curation, A.d.B.; writing—original draft preparation, A.d.B. and M.B.; writing—review and editing, A.d.B., A.D., F.M., J.-Y.B., I.R.-C. and M.B.; visualization, A.D.,

J.-Y.B., I.R.-C. and M.B.; supervision, M.B.; project administration, M.B.; funding acquisition, M.B. All authors read and agreed to the published version of the manuscript.

Funding: This research received no external funding.

Conflicts of Interest: J.-Y. Blay reports grants and personal fees from PharmaMar and AstraZeneca outside the submitted work. I. Ray-Coquard reports personal fees from Roche, PharmaMar, AstraZeneca, Clovis, GlaxoSmithKline, Bristol Myers Squibb, Agenus, Mersana, ImmunoGen, MSD, EISAI, and Novartis outside the submitted work. The other authors declare no conflict of interest.

References

1. Klemperer, P.; Rabin, C.B. Primary Neoplasms of the pleura. A report of five cases. *Am. J. Ind. Med.* **1992**, *22*, 4–31. [CrossRef] [PubMed]
2. Demicco, E.G.; Wagner, M.J.; Maki, R.G.; Gupta, V.; Iofin, I.; Lazar, A.; Wang, W.-L. Risk assessment in solitary fibrous tumors: Validation and refinement of a risk stratification model. *Mod. Pathol.* **2017**, *30*, 1433–1442. [CrossRef] [PubMed]
3. Diebold, M.; Soltermann, A.; Hottinger, S.; Haile, S.; Bubendorf, L.; Komminoth, P.; Jochum, W.; Grobholz, R.; Theegarten, D.; Berezowska, S.; et al. Prognostic value of MIB-1 proliferation index in solitary fibrous tumors of the pleura implemented in a new score—Amulticenter study. *Respir. Res.* **2017**, *18*, 210. [CrossRef] [PubMed]
4. Pasquali, S.; Gronchi, A.; Strauss, D.; Bonvalot, S.; Jeys, L.; Stacchiotti, S.; Hayes, A.; Honore, C.; Collini, P.; Renne, S.; et al. Resectable extra-pleural and extra-meningeal solitary fibrous tumours: A multi-centre prognostic study. *Eur. J. Surg. Oncol. (EJSO)* **2016**, *42*, 1064–1070. [CrossRef] [PubMed]
5. Salas, S.; Resseguier, N.; Blay, J.Y.; Le Cesne, A.; Italiano, A.; Chevreau, C.; Rosset, P.; Isambert, N.; Soulie, P.; Cupissol, D.; et al. Prediction of local and metastatic recurrence in solitary fibrous tumor: Construction of a risk calculator in a multicenter cohort from the French Sarcoma Group (FSG) database. *Ann. Oncol.* **2017**, *28*, 1779–1787. [CrossRef] [PubMed]
6. WHO. *Classification of Tumours Soft Tissue and Bone Tumours*, 5th ed.; IARC Press: Lyon, France, 2020.
7. de Pinieux, G.; Karanian, M.; Le Loarer, F.; Le Guellec, S.; Chabaud, S.; Terrier, P.; Bouvier, C.; Batistella, M.; Neuville, A.; Robin, Y.-M.; et al. Nationwide incidence of sarcomas and connective tissue tumors of intermediate malignancy over four years using an expert pathology review network. *PLoS ONE* **2021**, *16*, e0246958. [CrossRef]
8. Kinslow, C.J.; Bruce, S.S.; Rae, A.I.; Sheth, S.A.; McKhann, G.M.; Sisti, M.B.; Bruce, J.N.; Sonabend, A.M.; Wang, T.J.C. Solitary-fibrous tumor/hemangiopericytoma of the central nervous system: A population-based study. *J. Neuro-Oncol.* **2018**, *138*, 173–182. [CrossRef]
9. Kinslow, C.J.; Wang, T.J.C. Incidence of extrameningeal solitary fibrous tumors. *Cancer* **2020**, *126*, 4067. [CrossRef]
10. Kim, J.M.; Choi, Y.-L.; Kim, Y.J.; Park, H.K. Comparison and evaluation of risk factors for meningeal, pleural, and extrapleural solitary fibrous tumors: A clinicopathological study of 92 cases confirmed by STAT6 immunohistochemical staining. *Pathol.-Res. Pract.* **2017**, *213*, 619–625. [CrossRef]
11. Goodlad, J.R.; Fletcher, C.D.M. Solitary fibrous tumour arising at unusual sites: Analysis of a series. *Histopathology* **1991**, *19*, 515–522. [CrossRef]
12. Ronchi, A.; Cozzolino, I.; Zito Marino, F.; Accardo, M.; Montella, M.; Panarese, I.; Roccuzzo, G.; Toni, G.; Franco, R.; De Chiara, A. Extrapleural solitary fibrous tumor: A distinct entity from pleural solitary fibrous tumor. An update on clinical, molecular and diagnostic features. *Ann. Diagn. Pathol.* **2018**, *34*, 142–150. [CrossRef] [PubMed]
13. Gholami, S.; Cassidy, M.R.; Kirane, A.; Kuk, D.; Zanchelli, B.; Antonescu, C.R.; Singer, S.; Brennan, M. Size and Location are the Most Important Risk Factors for Malignant Behavior in Resected Solitary Fibrous Tumors. *Ann. Surg. Oncol.* **2017**, *24*, 3865–3871. [CrossRef] [PubMed]
14. Haas, R.L.; Walraven, I.; Lecointe-Artzner, E.; Van Houdt, W.J.; Strauss, D.; Schrage, Y.; Hayes, A.J.; Raut, C.P.; Fairweather, M.; Baldini, E.H.; et al. Extrameningeal solitary fibrous tumors—Surgery alone or surgery plus perioperative radiotherapy: A retrospective study from the global solitary fibrous tumor initiative in collaboration with the Sarcoma Patients EuroNet. *Cancer* **2020**, *126*, 3002–3012. [CrossRef] [PubMed]
15. Wignall, O.J.; Moskovic, E.C.; Thway, K.; Thomas, J.M. Solitary Fibrous Tumors of the Soft Tissues: Review of the Imaging and Clinical Features With Histopathologic Correlation. *Am. J. Roentgenol.* **2010**, *195*, W55–W62. [CrossRef]
16. Zafar, H.; Takimoto, C.H.; Weiss, G. Doege-Potter syndrome. *Med. Oncol.* **2003**, *20*, 403–407. [CrossRef]
17. Steigen, S.E.; Schaeffer, D.F.; West, R.B.; Nielsen, T.O. Expression of insulin-like growth factor 2 in mesenchymal neoplasms. *Mod. Pathol.* **2009**, *22*, 914–921. [CrossRef]
18. Le Roith, D. Tumor-Induced Hypoglycemia. *N. Engl. J. Med.* **1999**, *341*, 757–758. [CrossRef]
19. Davanzo, B.; Emerson, R.E.; Lisy, M.; Koniaris, L.G.; Kays, J.K. Solitary fibrous tumor. *Transl. Gastroenterol. Hepatol.* **2018**, *3*. [CrossRef]
20. Kallen, M.E.; Hornick, J.L. The 2020 WHO Classification: What's New in Soft Tissue Tumor Pathology? *Am. J. Surg. Pathol.* **2021**, *45*, e1–e23. [CrossRef]
21. Stout, A.P.; Murray, M.R. Hemangiopericytoma a Vascular Tumor Featuring Zimmermann's Pericytes. *Ann. Surg.* **1942**, *116*, 26–33. [CrossRef]

2. Fletcher, C.; Bridge, J.; Hogendoorn, P.; Mertens, F. *World Health Organization Classification of Tumours of Soft Tissue and Bone*, 4th ed.; IARC Press: Lyon, France, 2013.
3. Louis, D.N.; Perry, A.; Reifenberger, G.; von Deimlin, A.; Figarella-Branger, D.; Cavenee, W.K.; Ohgaki, H.; Wiestler, O.D.; Kleihues, P.; Ellison, D.W. The 2016 World Health Organization Classification of Tumors of the Central Nervous System: A summary. *Acta Neuropathol.* **2016**, *131*, 803–820. [CrossRef]
4. Louis, D.N.; Perry, A.; Wesseling, P.; Brat, D.J.; Cree, I.A.; Figarella-Branger, D.; Hawkins, C.; Ng, H.K.; Pfister, S.M.; Reifenberger, G.; et al. The 2021 WHO Classification of Tumors of the Central Nervous System: A summary. *Neuro-Oncol.* **2021**, *23*, 1231–1251. [CrossRef]
5. England, D.M.; Hochholzer, L.; McCarthy, M.J. Localized Benign and Malignant Fibrous Tumors of the Pleura: A Clinicopathologic Review of 223 Cases. *Am. J. Surg. Pathol.* **1989**, *13*, 640–658. [CrossRef]
6. de Perrot, M.; Fischer, S.; Bründler, M.-A.; Sekine, Y.; Keshavjee, S. Solitary fibrous tumors of the pleura. *Ann. Thorac. Surg.* **2002**, *74*, 285–293. [CrossRef]
7. Schirosi, L.; Lantuejoul, S.; Cavazza, A.; Murer, B.; Brichon, P.Y.; Migaldi, M.; Sartori, G.; Sgambato, A.; Rossi, G. Pleuro-pulmonary Solitary Fibrous Tumors: A Clinicopathologic, Immunohistochemical, and Molecular Study of 88 Cases Confirming the Prognostic Value of de Perrot Staging System and p53 Expression, and Evaluating the Role of c-kit, BRAF, PDGFRs (α/β), c-met, and EGFR. *Am. J. Surg. Pathol.* **2008**, *32*, 1627–1642.
8. Baldi, G.G.; Stacchiotti, S.; Mauro, V.; Tos, A.P.D.; Gronchi, A.; Pastorino, U.; Duranti, L.; Provenzano, S.; Marrari, A.; Libertini, M.; et al. Solitary fibrous tumor of all sites: Outcome of late recurrences in 14 patients. *Clin. Sarcoma Res.* **2013**, *3*, 4. [CrossRef]
9. Georgiesh, T.; Boye, K.; Bjerkehagen, B. A novel risk score to predict early and late recurrence in solitary fibrous tumour. *Histopathology* **2020**, *77*, 123–132. [CrossRef]
10. Mohamed, H.; Mandal, A.K. Natural History of Multifocal Solitary Fibrous Tumors of the Pleura: A 25-Year Follow-up Report. *J. Natl. Med. Assoc.* **2004**, *96*, 659–662.
11. Park, C.K.; Lee, D.H.; Park, J.Y.; Park, S.H.; Kwon, K.Y. Multiple Recurrent Malignant Solitary Fibrous Tumors: Long-Term Follow-Up of 24 Years. *Ann. Thorac. Surg.* **2011**, *91*, 1285–1288. [CrossRef]
12. Okike, N.; Bernatz, P.E.; Woolner, L.B. Localized mesothelioma of the pleura. *J. Thorac. Cardiovasc. Surg.* **1978**, *75*, 363–372. [CrossRef]
13. van Houdt, W.J.; Westerveld, C.M.A.; Vrijenhoek, J.E.P.; van Gorp, J.; van Coevorden, F.; Verhoef, C.; van Dalen, T. Prognosis of Solitary Fibrous Tumors: A Multicenter Study. *Ann. Surg. Oncol.* **2013**, *20*, 4090–4095. [CrossRef]
14. Wilky, B.A.; Montgomery, E.A.; Guzzetta, A.A.; Ahuja, N.; Meyer, C.F. Extrathoracic Location and "Borderline" Histology are Associated with Recurrence of Solitary Fibrous Tumors After Surgical Resection. *Ann. Surg. Oncol.* **2013**, *20*, 4080–4089. [CrossRef]
15. Cranshaw, I.; Gikas, P.; Fisher, C.; Thway, K.; Thomas, J.; Hayes, A. Clinical outcomes of extra-thoracic solitary fibrous tumours. *Eur. J. Surg. Oncol. (EJSO)* **2009**, *35*, 994–998. [CrossRef]
16. Reisenauer, J.S.; Mneimneh, W.; Jenkins, S.; Mansfield, A.S.; Aubry, M.C.; Fritchie, K.J.; Allen, M.S.; Blackmon, S.H.; Cassivi, S.D.; Nichols, F.C.; et al. Comparison of Risk Stratification Models to Predict Recurrence and Survival in Pleuropulmonary Solitary Fibrous Tumor. *J. Thorac. Oncol.* **2018**, *13*, 1349–1362. [CrossRef]
17. Demicco, E.G.; Park, M.S.; Araujo, D.M.; Fox, P.S.; Bassett, R.L.; Pollock, R.; Lazar, A.; Wang, W.-L. Solitary fibrous tumor: A clinicopathological study of 110 cases and proposed risk assessment model. *Mod. Pathol.* **2012**, *25*, 1298–1306. [CrossRef]
18. Mosquera, J.-M.; Fletcher, C.D.M. Expanding the Spectrum of Malignant Progression in Solitary Fibrous Tumors: A Study of 8 Cases With a Discrete Anaplastic Component—Is This Dedifferentiated SFT? *Am. J. Surg. Pathol.* **2009**, *33*, 1314–1321. [CrossRef]
19. Ghanim, B.; Hess, S.; Bertoglio, P.; Celik, A.; Bas, A.; Oberndorfer, F.; Melfi, F.; Mussi, A.; Klepetko, W.; Pirker, C.; et al. Intrathoracic solitary fibrous tumor—An international multicenter study on clinical outcome and novel circulating biomarkers. *Sci. Rep.* **2017**, *7*, 12557. [CrossRef]
20. Enzinger, F.M.; Smith, B.H. Hemangiopericytoma: An analysis of 106 cases. *Hum. Pathol.* **1976**, *7*, 61–82. [CrossRef]
21. Robinson, D.R.; Wu, Y.-M.; Kalyana-Sundaram, S.; Cao, X.; Lonigro, R.J.; Sung, Y.-S.; Chen, C.-L.; Zhang, L.; Wang, R.; Su, F.; et al. Identification of recurrent NAB2-STAT6 gene fusions in solitary fibrous tumor by integrative sequencing. *Nat. Genet.* **2013**, *45*, 180–185. [CrossRef]
22. Demicco, E.G.; Harms, P.; Patel, R.; Smith, S.C.; Ingram, D.; Torres, K.E.; Carskadon, S.L.; Camelo-Piragua, S.; McHugh, J.B.; Siddiqui, J.; et al. Extensive Survey of STAT6 Expression in a Large Series of Mesenchymal Tumors. *Am. J. Clin. Pathol.* **2015**, *143*, 672–682. [CrossRef]
23. Doyle, L.A.; Vivero, M.; Fletcher, C.D.M.; Mertens, F.; Hornick, J. Nuclear expression of STAT6 distinguishes solitary fibrous tumor from histologic mimics. *Mod. Pathol.* **2013**, *27*, 390–395. [CrossRef] [PubMed]
24. Koelsche, C.; Schweizer, L.; Renner, M.; Warth, A.; Jones, D.T.W.; Sahm, F.; Reuss, D.E.; Capper, D.; Knösel, T.; Schulz, B.; et al. Nuclear relocation of STAT6 reliably predicts NAB2-STAT6 fusion for the diagnosis of solitary fibrous tumour. *Histopathology* **2014**, *65*, 613–622. [CrossRef] [PubMed]
25. Schweizer, L.; Koelsche, C.; Sahm, F.; Piro, R.M.; Capper, D.; Reuss, D.E.; Pusch, S.; Habel, A.; Meyer, J.; Göck, T.; et al. Meningeal hemangiopericytoma and solitary fibrous tumors carry the NAB2-STAT6 fusion and can be diagnosed by nuclear expression of STAT6 protein. *Acta Neuropathol.* **2013**, *125*, 651–658. [CrossRef]
26. Yoshida, A.; Tsuta, K.; Ohno, M.; Yoshida, M.; Narita, Y.; Kawai, A.; Asamura, H.; Kushima, R. STAT6 Immunohistochemistry Is Helpful in the Diagnosis of Solitary Fibrous Tumors. *Am. J. Surg. Pathol.* **2014**, *38*, 552–559. [CrossRef]

47. Doyle, L.A.; Tao, D.; Mariño-Enríquez, A. STAT6 is amplified in a subset of dedifferentiated liposarcoma. *Mod. Pathol.* **2014**, *2?*, 1231–1237. [CrossRef]
48. Bouvier, C.; Bertucci, F.; Métellus, P.; Finetti, P.; De Paula, A.M.; Forest, F.; Mokhtari, K.; Miquel, C.; Birnbaum, D.; Vasiljevi? A.; et al. ALDH1 is an immunohistochemical diagnostic marker for solitary fibrous tumours and haemangiopericytomas of th? meninges emerging from gene profiling study. *Acta Neuropathol. Commun.* **2013**, *1*, 10. [CrossRef]
49. Vivero, M.; Doyle, L.A.; Fletcher, C.D.M.; Mertens, F.; Hornick, J.L. GRIA2 is a novel diagnostic marker for solitary fibrous tumou? identified through gene expression profiling. *Histopathology* **2014**, *65*, 71–80. [CrossRef]
50. Barthelmeß, S.; Geddert, H.; Boltze, C.; Moskalev, E.A.; Bieg, M.; Sirbu, H.; Brors, B.; Wiemann, S.; Hartmann, A.; Agaimy, A.; et a? Solitary Fibrous Tumors/Hemangiopericytomas with Different Variants of the NAB2-STAT6 Gene Fusion Are Characterized b? Specific Histomorphology and Distinct Clinicopathological Features. *Am. J. Pathol.* **2014**, *184*, 1209–1218. [CrossRef]
51. Chmielecki, J.; Crago, A.; Rosenberg, M.; O'Connor, R.; Walker, S.R.; Ambrogio, L.; Auclair, D.; McKenna, A.; Heinrich, M.; Frank D.A.; et al. Whole-exome sequencing identifies a recurrent NAB2-STAT6 fusion in solitary fibrous tumors. *Nat. Genet.* **2013**, *4?*, 131–132. [CrossRef]
52. Mohajeri, A.; Tayebwa, J.; Collin, A.; Nilsson, J.; Magnusson, L.; von Steyern, F.V.; Brosjö, O.; Domanski, H.A.; Larsson, O.; Scio? R.; et al. Comprehensive genetic analysis identifies a pathognomonic NAB2/STAT6 fusion gene, nonrandom secondary genomi? imbalances, and a characteristic gene expression profile in solitary fibrous tumor. *Genes Chromosomes Cancer* **2013**, *52*, 873–88? [CrossRef]
53. Schwachtgen, J.L.; Houston, P.; Campbell, C.; Sukhatme, V.; Braddock, M. Fluid shear stress activation of egr-1 transcription i? cultured human endothelial and epithelial cells is mediated via the extracellular signal-related kinase 1/2 mitogen-activate? protein kinase pathway. *J. Clin. Invest.* **1998**, *101*, 2540–2549. [CrossRef]
54. Cheng, J.-C.; Chang, H.-M.; Leung, P.C.K. Egr-1 mediates epidermal growth factor-induced downregulation of E-cadheri? expression via Slug in human ovarian cancer cells. *Oncogene* **2012**, *32*, 1041–1049. [CrossRef]
55. Ji, R.-C. Hypoxia and lymphangiogenesis in tumor microenvironment and metastasis. *Cancer Lett.* **2014**, *346*, 6–16. [CrossRef]
56. Binnemars-Postma, K.; Bansal, R.; Storm, G.; Prakash, J. Targeting the Stat6 pathway in tumor-associated macrophages reduce? tumor growth and metastatic niche formation in breast cancer. *FASEB J.* **2018**, *32*, 969–978. [CrossRef]
57. Shahmarvand, N.; Nagy, A.; Shahryari, J.; Ohgami, R.S. Mutations in the signal transducer and activator of transcription famil? of genes in cancer. *Cancer Sci.* **2018**, *109*, 926–933. [CrossRef]
58. Li, J.; Rodriguez, J.P.; Niu, F.; Pu, M.; Wang, J.; Hung, L.-W.; Shao, Q.; Zhu, Y.; Ding, W.; Liu, Y.; et al. Structural basis for DNA recognition by STAT6. *Proc. Natl. Acad. Sci. USA* **2016**, *113*, 13015–13020. [CrossRef]
59. Akaike, K.; Kurisaki-Arakawa, A.; Hara, K.; Suehara, Y.; Takagi, T.; Mitani, K.; Kaneko, K.; Yao, T.; Saito, T. Distinct clinico? pathological features of NAB2-STAT6 fusion gene variants in solitary fibrous tumor with emphasis on the acquisition of highl? malignant potential. *Hum. Pathol.* **2014**, *46*, 347–356. [CrossRef]
60. Yuzawa, S.; Nishihara, H.; Wang, L.; Tsuda, M.; Kimura, T.; Tanino, M.; Tanaka, S. Analysis of NAB2-STAT6 Gene Fusion i? 17 Cases of Meningeal Solitary Fibrous Tumor/Hemangiopericytoma: Review of the Literature. *Am. J. Surg. Pathol.* **2016**, *40*, 1031–1040. [CrossRef]
61. Chuang, I.-C.; Liao, K.-C.; Huang, H.-Y.; Kao, Y.-C.; Li, C.-F.; Huang, S.-C.; Tsai, J.-W.; Chen, K.-C.; Lan, J.; Lin, P.-C. NAB2-STAT? gene fusion and STAT6 immunoexpression in extrathoracic solitary fibrous tumors: The association between fusion variants an? locations: Extrathoracic solitary fibrous tumor. *Pathol. Int.* **2016**, *66*, 288–296. [CrossRef]
62. Huang, S.; Li, C.; Kao, Y.; Chuang, I.; Tai, H.; Tsai, J.; Yu, S.; Huang, H.; Lan, J.; Yen, S.; et al. The clinicopathological significance o? NAB 2- STAT 6 gene fusions in 52 cases of intrathoracic solitary fibrous tumors. *Cancer Med.* **2015**, *5*, 159–168. [CrossRef]
63. Park, Y.-S.; Kim, H.-S.; Kim, J.-H.; Choi, S.-H.; Kim, D.-S.; Ryoo, Z.Y.; Kim, J.-Y.; Lee, S. NAB2-STAT6 fusion protein mediates cel? proliferation and oncogenic progression via EGR-1 regulation. *Biochem. Biophys. Res. Commun.* **2020**, *526*, 287–292. [CrossRef]
64. Georgiesh, T.; Namløs, H.M.; Sharma, N.; Lorenz, S.; Myklebost, O.; Bjerkehagen, B.; Meza-Zepeda, L.A.; Boye, K. Clinical and molecular implications of NAB2-STAT6 fusion variants in solitary fibrous tumour. *Pathology* **2021**, *53*, 713–719. [CrossRef]
65. Casali, P.G.; Abecassis, N.; Bauer, S.; Biagini, R.; Bielack, S.; Bonvalot, S.; Boukovinas, I.; Bovee, J.V.M.G.; Brodowicz, T.; Broto, J.M.; et al. Soft tissue and visceral sarcomas: ESMO–EURACAN Clinical Practice Guidelines for diagnosis, treatment and follow-up? *Ann. Oncol.* **2018**, *29*, iv51–iv67. [CrossRef]
66. Casali, P.G.; Bielack, S.; Abecassis, N.; Aro, H.T.; Bauer, S.; Biagini, R.; Bonvalot, S.; Boukovinas, I.; Bovee, J.V.M.G.; Brennan, B.; et al. Bone sarcomas: ESMO–PaedCan–EURACAN Clinical Practice Guidelines for diagnosis, treatment and follow-up. *Ann. Oncol.* **2018**, *29*, iv79–iv95. [CrossRef]
67. Casali, P.G.; Blay, J.Y.; Abecassis, N.; Bajpai, J.; Bauer, S.; Biagini, R.; Bielack, S.; Bonvalot, S.; Boukovinas, I.; Bovee, J.V.M.G.; et al. Gastrointestinal stromal tumours: ESMO–EURACAN–GENTURIS Clinical Practice Guidelines for diagnosis, treatment and follow-up. *Ann. Oncol.* **2022**, *33*, 20–33. [CrossRef]
68. Blay, J.-Y.; Honoré, C.; Stoeckle, E.; Meeus, P.; Jafari, M.; Gouin, F.; Anract, P.; Ferron, G.; Rochwerger, A.; Ropars, M.; et al. Surgery in reference centers improves survival of sarcoma patients: A nationwide study. *Ann. Oncol.* **2019**, *30*, 1143–1153. [CrossRef]
69. Benjamin, R.S.; Choi, H.; Macapinlac, H.A.; Burgess, M.A.; Patel, S.R.; Chen, L.L.; Podoloff, D.A.; Charnsangavej, C. We Should Desist Using RECIST, at Least in GIST. *JCO* **2007**, *25*, 1760–1764. [CrossRef]
70. Choi, H.; Charnsangavej, C.; Faria, S.C.; Macapinlac, H.A.; Burgess, M.A.; Patel, S.R.; Chen, L.L.; Podoloff, D.A.; Benjamin, R.S. Correlation of Computed Tomography and Positron Emission Tomography in Patients With Metastatic Gastrointestinal Stromal

Tumor Treated at a Single Institution With Imatinib Mesylate: Proposal of New Computed Tomography Response Criteria. *J. Clin. Oncol.* **2007**, *25*, 1753–1759. [CrossRef]

71. Stacchiotti, S.; Verderio, P.; Messina, A.; Morosi, C.; Collini, P.; Llombart-Bosch, A.; Martin, J.; Comandone, A.; Cruz, J.; Ferraro, A.; et al. Tumor response assessment by modified Choi criteria in localized high-risk soft tissue sarcoma treated with chemotherapy. *Cancer* **2012**, *118*, 5857–5866. [CrossRef]

72. Park, M.S.; Ravi, V.; Conley, A.; Patel, S.R.; Trent, J.C.; Lev, D.C.; Lazar, A.J.; Wang, W.-L.; Benjamin, R.S.; Araujo, D.M. The role of chemotherapy in advanced solitary fibrous tumors: A retrospective analysis. *Clin. Sarcoma Res.* **2013**, *3*, 7. [CrossRef]

73. Park, M.S.; Patel, S.R.; Ludwig, J.A.; Trent, J.C.; Conrad, C.A.; Lazar, A.J.; Wang, W.-L.; Boonsirikamchai, P.; Choi, H.; Wang, X.; et al. Activity of temozolomide and bevacizumab in the treatment of locally advanced, recurrent, and metastatic hemangiopericytoma and malignant solitary fibrous tumor: Temozolomide/Bevacizumab Therapy in HPC/SFT. *Cancer* **2011**, *117*, 4939–4947. [CrossRef]

74. Stacchiotti, S.; Libertini, M.; Negri, T.; Palassini, E.; Gronchi, A.; Fatigoni, S.; Poletti, P.; Vincenzi, B.; Tos, A.D.; Mariani, L.; et al. Response to chemotherapy of solitary fibrous tumour: A retrospective study. *Eur. J. Cancer* **2013**, *49*, 2376–2383. [CrossRef]

75. Stacchiotti, S.; Negri, T.; Libertini, M.; Palassini, E.; Marrari, A.; De Troia, B.; Gronchi, A.; Tos, A.D.; Morosi, C.; Messina, A.; et al. Sunitinib malate in solitary fibrous tumor (SFT). *Ann. Oncol.* **2012**, *23*, 3171–3179. [CrossRef]

76. Maruzzo, M.; Martin-Liberal, J.; Messiou, C.; Miah, A.; Thway, K.; Alvarado, R.; Judson, I.; Benson, C. Pazopanib as first line treatment for solitary fibrous tumours: The Royal Marsden Hospital experience. *Clin. Sarcoma Res.* **2015**, *5*, 5. [CrossRef]

77. Cardillo, G.; Lococo, F.; Carleo, F.; Martelli, M. Solitary fibrous tumors of the pleura. *Curr. Opin. Pulm. Med.* **2012**, *18*, 339–346. [CrossRef]

78. Espat, N.J.; Lewis, J.J.; Leung, D.; Woodruff, J.M.; Antonescu, C.R.; Shia, J.; Brennan, M.F. Conventional hemangiopericytoma: Modern analysis of outcome. *Cancer* **2002**, *95*, 1746–1751. [CrossRef]

79. Robinson, L.A. Solitary Fibrous Tumor of the Pleura. *Cancer Control.* **2006**, *13*, 6. [CrossRef]

80. Spitz, F.R.; Bouvet, M.; Pisters, P.W.T.; Pollock, R.E.; Feig, B.W. Hemangiopericytoma: A 20-year single-institution experience. *Ann. Surg. Oncol.* **1998**, *5*, 350–355. [CrossRef]

81. Lahon, B.; Mercier, O.; Fadel, E.; Ghigna, M.R.; Petkova, B.; Mussot, S.; Fabre, D.; Le Chevalier, T.; Dartevelle, P. Solitary Fibrous Tumor of the Pleura: Outcomes of 157 Complete Resections in a Single Center. *Ann. Thorac. Surg.* **2012**, *94*, 394–400. [CrossRef]

82. Nomori, H. Contacting metastasis of a fibrous tumor of the pleura. *Eur. J. Cardio-Thoracic Surg.* **1997**, *12*, 928–930. [CrossRef]

83. Lococo, F.; Cesario, A.; Cardillo, G.; Filosso, P.L.; Galetta, D.; Carbone, L.; Oliaro, A.; Spaggiari, L.; Cusumano, G.; Margaritora, S.; et al. Malignant Solitary Fibrous Tumors of the Pleura: Retrospective Review of a Multicenter Series. *J. Thorac. Oncol.* **2012**, *7*, 1698–1706. [CrossRef] [PubMed]

84. Bowe, S.N.; Wakely, P.E., Jr.; Ozer, E. Head and neck solitary fibrous tumors: Diagnostic and therapeutic challenges. *Laryngoscope* **2012**, *122*, 1748–1755. [CrossRef] [PubMed]

85. Wang, X.; Qian, J.; Bi, Y.; Ping, B.; Zhang, R. Malignant transformation of orbital solitary fibrous tumor. *Int. Ophthalmol.* **2013**, *33*, 299–303. [CrossRef] [PubMed]

86. Yang, X.; Zheng, J.; Ye, W.; Wang, Y.; Zhu, H.; Wang, L.; Zhang, Z. Malignant solitary fibrous tumors of the head and neck: A clinicopathological study of nine consecutive patients. *Oral Oncol.* **2009**, *45*, 678–682. [CrossRef]

87. Wushou, A.; Jiang, Y.-Z.; Liu, Y.-R.; Shao, Z.-M. The demographic features, clinicopathologic characteristics, treatment outcome and disease-specific prognostic factors of solitary fibrous tumor: A population-based analysis. *Oncotarget* **2015**, *6*, 41875–41883. [CrossRef]

88. Bishop, A.J.; Zagars, G.K.; Demicco, E.G.; Wang, W.-L.; Feig, B.W.; Guadagnolo, B.A. Soft Tissue Solitary Fibrous Tumor: Combined Surgery and Radiation Therapy Results in Excellent Local Control. *Am. J. Clin. Oncol.* **2018**, *41*, 81–85. [CrossRef]

89. Bonvalot, S.; Gronchi, A.; Le Péchoux, C.; Swallow, C.J.; Strauss, D.; Meeus, P.; van Coevorden, F.; Stoldt, S.; Stoeckle, E.; Rutkowski, P.; et al. Preoperative radiotherapy plus surgery versus surgery alone for patients with primary retroperitoneal sarcoma (EORTC-62092: STRASS): A multicentre, open-label, randomised, phase 3 trial. *Lancet Oncol.* **2020**, *21*, 1366–1377. [CrossRef]

90. Haas, R.L.; Walraven, I.; Lecointe-Artzner, E.; Scholten, A.N.; van Houdt, W.; Griffin, A.M.; Ferguson, P.C.; Miah, A.B.; Zaidi, S.; DeLaney, T.F.; et al. Radiation Therapy as Sole Management for Solitary Fibrous Tumors (SFT): A Retrospective Study From the Global SFT Initiative in Collaboration With the Sarcoma Patients EuroNet. *Int. J. Radiat. Oncol.* **2018**, *101*, 1226–1233. [CrossRef]

91. Oike, N.; Kawashima, H.; Ogose, A.; Hotta, T.; Hirano, T.; Ariizumi, T.; Yamagishi, T.; Umezu, H.; Inagawa, S.; Endo, N. A malignant solitary fibrous tumour arising from the first lumbar vertebra and mimicking an osteosarcoma: A case report. *World J. Surg. Oncol.* **2017**, *15*, 100. [CrossRef]

92. De Boer, J.; Jager, P.L.; Wiggers, T.; Nieboer, P.; Wymenga, A.N.M.; Pras, E.; Hoogenberg, K.; Sleijfer, D.T.; Suurmeijer, A.J.; Van Der Graaf, W.T. The therapeutic challenge of a nonresectable solitary fibrous tumor in a hypoglycemic patient. *Int. J. Clin. Oncol.* **2006**, *11*, 478–481. [CrossRef]

93. Han, G.; Zhang, Z.; Shen, X.; Wang, K.; Zhao, Y.; He, J.; Gao, Y.; Shan, X.; Xin, G.; Li, C.; et al. Doege-Potter syndrome: A review of the literature including a new case report. *Medicine* **2017**, *96*, e7417. [CrossRef] [PubMed]

94. Constantinidou, A.; Jones, R.L.; Olmos, D.; Thway, K.; Fisher, C.; Al-Muderis, O.; Judson, I. Conventional anthracycline-based chemotherapy has limited efficacy in solitary fibrous tumour. *Acta Oncol.* **2011**, *51*, 550–554. [CrossRef] [PubMed]

95. Levard, A.; Derbel, O.; Méeus, P.; Ranchère, D.; Ray-Coquard, I.; Blay, J.-Y.; Cassier, P.A. Outcome of patients with advanced solitary fibrous tumors: The Centre Léon Bérard experience. *BMC Cancer* **2013**, *13*, 109. [CrossRef]

96. Stacchiotti, S.; Tortoreto, M.; Bozzi, F.; Tamborini, E.; Morosi, C.; Messina, A.; Libertini, M.; Palassini, E.; Cominetti, D.; Negri, T.; et al. Dacarbazine in Solitary Fibrous Tumor: A Case Series Analysis and Preclinical Evidence vis-à-vis Temozolomide and Antiangiogenics. *Clin. Cancer Res.* **2013**, *19*, 5192–5201. [CrossRef] [PubMed]

97. Khalifa, J.; Ouali, M.; Chaltiel, L.; Le Guellec, S.; Le Cesne, A.; Blay, J.-Y.; Cousin, P.; Chaigneau, L.; Bompas, E.; Piperno-Neumann, S.; et al. Efficacy of trabectedin in malignant solitary fibrous tumors: A retrospective analysis from the French Sarcoma Group. *BMC Cancer* **2015**, *15*, 1–8. [CrossRef] [PubMed]

98. Le Cesne, A.; Ray-Coquard, I.; Duffaud, F.; Chevreau, C.; Penel, N.; Nguyen, B.B.; Piperno-Neumann, S.; Delcambre, C.; Rios, M.; Chaigneau, L.; et al. Trabectedin in patients with advanced soft tissue sarcoma: A retrospective national analysis of the French Sarcoma Group. *Eur. J. Cancer* **2015**, *51*, 742–750. [CrossRef] [PubMed]

99. Schöffski, P.; Timmermans, I.; Hompes, D.; Stas, M.; Sinnaeve, F.; De Leyn, P.; Coosemans, W.; Van Raemdonck, D.; Hauben, E.; Sciot, R.; et al. Clinical Presentation, Natural History, and Therapeutic Approach in Patients with Solitary Fibrous Tumor: A Retrospective Analysis. *Sarcoma* **2020**, *2020*, 1–9. [CrossRef]

100. Outani, H.; Kobayashi, E.; Wasa, J.; Saito, M.; Takenaka, S.; Hayakawa, K.; Endo, M.; Takeuchi, A.; Kobayashi, H.; Kito, M.; et al. Clinical Outcomes of Patients with Metastatic Solitary Fibrous Tumors: A Japanese Musculoskeletal Oncology Group (JMOG) Multiinstitutional Study. *Ann. Surg. Oncol.* **2020**, *28*, 3893–3901. [CrossRef]

101. Mulamalla, K.; Truskinovsky, A.M.; Dudek, A.Z. Rare case of hemangiopericytoma responds to sunitinib. *Transl. Res.* **2008**, *151*, 129–133. [CrossRef]

102. De Pas, T.; Toffalorio, F.; Colombo, P.; Trifirò, G.; Pelosi, G.; Della Vigna, P.; Manzotti, M.; Agostini, M.; De Braud, F.G.M. Brief Report: Activity of Imatinib in a Patient with Platelet-Derived-Growth-Factor Receptor Positive Malignant Solitary Fibrous Tumor of the Pleura. *J. Thorac. Oncol.* **2008**, *3*, 938–941. [CrossRef]

103. George, S.; Merriam, P.; Maki, R.G.; Van den Abbeele, A.D.; Yap, J.T.; Akhurst, T.; Harmon, D.C.; Bhuchar, G.; O'Mara, M.M.; D'Adamo, D.R.; et al. Multicenter Phase II Trial of Sunitinib in the Treatment of Nongastrointestinal Stromal Tumor Sarcomas. *JCO* **2009**, *27*, 3154–3160. [CrossRef] [PubMed]

104. Domont, J.; Massard, C.; Lassau, N.; Armand, J.-P.; Le Cesne, A.; Soria, J.-C. Hemangiopericytoma and antiangiogenic therapy: Clinical benefit of antiangiogenic therapy (sorafenib and sunitinib) in relapsed Malignant Haemangioperyctoma /Solitary Fibrous Tumour. *Investig. New Drugs* **2010**, *28*, 199–202. [CrossRef] [PubMed]

105. Valentin, T.; Fournier, C.; Penel, N.; Bompas, E.; Chaigneau, L.; Isambert, N.; Chevreau, C. Sorafenib in patients with progressive malignant solitary fibrous tumors: A subgroup analysis from a phase II study of the French Sarcoma Group (GSF/GETO). *Investig. New Drugs* **2013**, *31*, 1626–1627. [CrossRef] [PubMed]

106. Stacchiotti, S.; Negri, T.; Palassini, E.; Conca, E.; Gronchi, A.; Morosi, C.; Messina, A.; Pastorino, U.; Pierotti, M.A.; Casali, P.G.; et al. Sunitinib Malate and Figitumumab in Solitary Fibrous Tumor: Patterns and Molecular Bases of Tumor Response. *Mol. Cancer Ther.* **2010**, *9*, 1286–1297. [CrossRef]

107. Stacchiotti, S.; Tortoreto, M.; Baldi, G.; Grignani, G.; Toss, A.; Badalamenti, G.; Cominetti, D.; Morosi, C.; Tos, A.D.; Festinese, F.; et al. Preclinical and clinical evidence of activity of pazopanib in solitary fibrous tumour. *Eur. J. Cancer* **2014**, *50*, 3021–3028. [CrossRef]

108. Martin-Broto, J.; Stacchiotti, S.; Lopez-Pousa, A.; Redondo, A.; Bernabeu, D.; de Alava, E.; Casali, P.G.; Italiano, A.; Gutierrez, A.; Moura, D.S.; et al. Pazopanib for treatment of advanced malignant and dedifferentiated solitary fibrous tumour: A multicentre, single-arm, phase 2 trial. *Lancet Oncol.* **2019**, *20*, 134–144. [CrossRef]

109. Martin-Broto, J.; Cruz, J.; Penel, N.; Le Cesne, A.; Hindi, N.; Luna, P.; Moura, D.S.; Bernabeu, D.; de Alava, E.; Lopez-Guerrero, J.A.; et al. Pazopanib for treatment of typical solitary fibrous tumours: A multicentre, single-arm, phase 2 trial. *Lancet Oncol.* **2020**, *21*, 456–466. [CrossRef]

110. Kobayashi, H.; Iwata, S.; Wakamatsu, T.; Hayakawa, K.; Yonemoto, T.; Wasa, J.; Oka, H.; Ueda, T.; Tanaka, S. Efficacy and safety of trabectedin for patients with unresectable and relapsed soft-tissue sarcoma in Japan: A Japanese Musculoskeletal Oncology Group study. *Cancer* **2019**, *126*, 1253–1263. [CrossRef]

111. Hatva, E. Vascular Growth Factors and Receptors in Capillary Hemangioblastomas and Hemang iopericytomas. *Am. J. Pathol.* **1996**, *148*, 13.

112. Quek, R.; Wang, Q.; Morgan, J.A.; Shapiro, G.I.; Butrynski, J.E.; Ramaiya, N.; Huftalen, T.; Jederlinic, N.; Manola, J.; Wagner, A.J.; et al. Combination mTOR and IGF-1R Inhibition: Phase I Trial of Everolimus and Figitumumab in Patients with Advanced Sarcomas and Other Solid Tumors. *Clin. Cancer Res.* **2010**, *17*, 871–879. [CrossRef]

113. Tazzari, M.; Negri, T.; Rini, F.; Vergani, B.; Huber, V.; Villa, A.; Dagrada, P.; Colombo, C.; Fiore, M.; Gronchi, A.; et al. Adaptive immune contexture at the tumour site and downmodulation of circulating myeloid-derived suppressor cells in the response of solitary fibrous tumour patients to anti-angiogenic therapy. *Br. J. Cancer* **2014**, *111*, 1350–1362. [CrossRef] [PubMed]

114. Kamamoto, D.; Ohara, K.; Kitamura, Y.; Yoshida, K.; Kawakami, Y.; Sasaki, H. Association between programmed cell death ligand-1 expression and extracranial metastasis in intracranial solitary fibrous tumor/hemangiopericytoma. *J. Neuro-Oncol.* **2018**, *139*, 251–259. [CrossRef] [PubMed]

115. Dancsok, A.R.; Setsu, N.; Gao, D.; Blay, J.-Y.; Thomas, D.; Maki, R.G.; Nielsen, T.O.; Demicco, E.G. Expression of lymphocyte immunoregulatory biomarkers in bone and soft-tissue sarcomas. *Mod. Pathol.* **2019**, *32*, 1772–1785. [CrossRef] [PubMed]

116. Berghoff, A.S.; Kresl, P.; Rajky, O.; Widhalm, G.; Ricken, G.; Hainfellner, J.A.; Marosi, C.; Birner, P.; Preusser, M. Analysis of the inflammatory tumor microenvironment in meningeal neoplasms. *Clin. Neuropathol.* **2020**, *39*, 256–262. [CrossRef]

17. Boothe, J.T.; Budd, G.T.; Smolkin, M.B.; Ma, P.C. Durable Near-Complete Response to Anti-PD-1 Checkpoint Immunotherapy in a Refractory Malignant Solitary Fibrous Tumor of the Pleura. *Case Rep. Oncol.* **2017**, *10*, 998–1005. [CrossRef]
18. Toulmonde, M.; Penel, N.; Adam, J.; Chevreau, C.; Blay, J.-Y.; Le Cesne, A.; Bompas, E.; Piperno-Neumann, S.; Cousin, S.; Grellety, T.; et al. Use of PD-1 Targeting, Macrophage Infiltration, and IDO Pathway Activation in Sarcomas: A Phase 2 Clinical Trial. *JAMA Oncol.* **2018**, *4*, 93. [CrossRef]
19. Tap, W.D.; Wagner, A.J.; Schöffski, P.; Martin-Broto, J.; Krarup-Hansen, A.; Ganjoo, K.N.; Yen, C.-C.; Abdul Razak, A.R.; Spira, A.; Kawai, A.; et al. Effect of Doxorubicin Plus Olaratumab vs Doxorubicin Plus Placebo on Survival in Patients With Advanced Soft Tissue Sarcomas: The ANNOUNCE Randomized Clinical Trial. *JAMA* **2020**, *323*, 1266. [CrossRef]

Article

Clinical, Histological, and Molecular Features of Solitary Fibrous Tumor of Bone: A Single Institution Retrospective Review

Giuseppe Bianchi [1], Debora Lana [1], Marco Gambarotti [2], Cristina Ferrari [3], Marta Sbaraglia [4], Elena Pedrini [5], Laura Pazzaglia [3], Luca Sangiorgi [5], Isabella Bartolotti [5], Angelo Paolo Dei Tos [6], Katia Scotlandi [3] and Alberto Righi [2,*]

[1] Department of Orthopedic Oncology, IRCCS Istituto Ortopedico Rizzoli, 40136 Bologna, Italy; giuseppe.bianchi@ior.it (G.B.); debora.lana88@gmail.com (D.L.)
[2] Department of Pathology, IRCCS Istituto Ortopedico Rizzoli, 40136 Bologna, Italy; marco.gambarotti@ior.it
[3] Experimental Oncology Laboratory, IRCCS Istituto Ortopedico Rizzoli, 40136 Bologna, Italy; cristina.ferrari@ior.it (C.F.); laura.pazzaglia@ior.it (L.P.); katia.scotlandi@ior.it (K.S.)
[4] Department of Pathology, Azienda Ospedaliera di Padova, 35121 Padua, Italy; marta.sbaraglia@aopd.veneto.it
[5] Department of Rare Skeletal Disorders, IRCCS Istituto Ortopedico Rizzoli, 40136 Bologna, Italy; elena.pedrini@ior.it (E.P.); luca.sangiorgi@ior.it (L.S.); isabella.bartolotti@ior.it (I.B.)
[6] Department of Medicine, University of Padua School of Medicine, 35121 Padua, Italy; angelo.deitos@unipd.it
* Correspondence: alberto.righi@ior.it; Tel.: +39-051-636-6665

Citation: Bianchi, G.; Lana, D.; Gambarotti, M.; Ferrari, C.; Sbaraglia, M.; Pedrini, E.; Pazzaglia, L.; Sangiorgi, L.; Bartolotti, I.; Dei Tos, A.P.; et al. Clinical, Histological, and Molecular Features of Solitary Fibrous Tumor of Bone: A Single Institution Retrospective Review. *Cancers* **2021**, *13*, 2470. https://doi.org/10.3390/cancers13102470

Academic Editor: Bahil Ghanim

Received: 29 April 2021
Accepted: 16 May 2021
Published: 19 May 2021

Publisher's Note: MDPI stays neutral with regard to jurisdictional claims in published maps and institutional affiliations.

Simple Summary: Solitary fibrous tumors arising from the bone are an extremely rare event and only few cases have been previously described in the literature. It is characterized by a prominent branched vascularization, with a thin and dilated vascular texture defined as "staghorn" and by the presence of the *NAB2-STAT6* gene rearrangement, present in about 90% of cases and considered a pathognomonic feature. In the present study, we described our series of 24 cases of primary solitary fibrous tumor of the bone to find any clinical and molecular prognostic factors and to compare them with those currently used for soft tissue solitary fibrous tumor and to evaluate the risk stratification system proposed by Demicco, in order to understand whether this system was able to correctly predict the risk of local and distant metastatic relapse even in the case of solitary fibrous tumor of the bone.

Abstract: Primary solitary fibrous tumor (SFT) of the bone is extremely rare, with only few cases reported in the literature. We retrieved all cases of primary SFT of the bone treated at our institution and we assessed the morphology and the immunohistochemical and molecular features to investigate the clinical outcome of primary SFT of the bone and any clinical relevance of clinical and histological criteria of aggressiveness currently adopted for the soft tissues counterpart. Morphologically, 15 cases evidenced high cellularity, cytologic atypia, and foci of necrosis and were associated with more than 4 mitotic figures/10 HPF. Immunohistochemical analysis showed an expression of CD34 and of STAT6 immunopositivity in 95% and in 100% of cases, respectively. The presence of *NAB2-STAT6* chimeric transcripts was found in 10 out of 12 cases in which RT-PCR analysis was feasible, whereas *TERT* promoter mutations analysis was feasible in 16 cases and only a C-to-T substitution in a heterozygous state was found in one DNA sample for the C228T genetic variant. *P53* variants were assessed in 12 cases: 11 (91.6%) cases showed a variation, while in one case, no alteration was found. Disease-specific survival was 64% at 5 years and 49% at 10 years. Statistical analysis showed no correlation between survival and all the clinicopathological and molecular parameters evaluated. In conclusion, at difference to SFT of soft tissues, aggressive behavior of primary SFT of the bone seems to be independent from mitotic count or any other clinicopathological and molecular features.

Keywords: solitary fibrous tumor; primary bone tumor; risk stratification; prognosis; *NAB2-STAT6* fusion transcripts

---

1. Introduction

Solitary fibrous tumor (SFT) is a rare mesenchymal tumor of fibroblastic origin that can occur at any anatomic site and typically affects middle-aged adults [1–3]. It is characterized by a strong morphologic heterogeneity with a wide spectrum of biologic features. The histological and molecular diagnostic criteria used in soft tissue SFT (S-SFT) have been recently applied on "non otherwise classified" primary bone tumors, drawing out a new category of SFT of the bone (B-SFT) [4–7]. Nevertheless, B-SFT is exceedingly rare, with only few cases are described in the literature [8–11], and its biological behavior has not yet been assessed. From a histopathological and molecular point of view, primary B-SFT shares the same features of S-SFT. It is characterized by a prominent, branched vascularization, with a thin and dilated vascular texture defined as "staghorn" and by the presence of the *NAB2-STAT6* gene rearrangement (NGFI-A binding protein 2—Signal Transducer and Activator of Transcription 6), present in about 90% of cases and considered a pathognomonic feature [1,12]. Positivity to CD34 stain is distinctive in 90–95% of the cases. S-SFT has an intermediate malignant potential with a low risk of metastasis. Some studies have investigated the prognostic role of previously described molecular markers, without, however, obtaining conclusive results; the aforementioned prognostic criteria have never been explored in B-SFT [13–16]. Most S-SFTs are clinically indolent, with an intermediate malignant potential and a low risk of metastasis, showing an overall 5- and 10-year distant metastasis (DM)-free rates of 74% and 55%, respectively. In recent times, different stratification risk models have been proposed [17–22]. The current most utilized scoring system to discriminate different risk groups for S-SFT—also related to the development of distal metastasis—is the one proposed by Demicco et al. [22], which considers patient age, mitotic activity, tumor necrosis, and size. To date, few prognostic molecular markers have been analyzed. *NAB2–STAT6* chimeric transcripts, with a frequency ranging from 55 to 100% [23,24], and characterized by different breakpoints in fusion genes, might contribute to the morphologic diversity of SFT; some studies evidenced associations between specific fusion variants and different clinical features [21,25]. In addition, specific point mutations within the promoter region of telomerase reverse transcriptase (*TERT*)—C228Tand C250T—have been recently reported in S-SFT subsets and other tumors [15,25–27]. These mutations confer enhanced *TERT* promoter activity and have been suggested as predictive factors to identify high-risk patients. Finally, TP53 has also been proposed as an SFTs risk factor. In particular, tumors with TP53 mutations were almost always classified as high risk [21,28]. Due to the rarity of B-SFT and taking advantage of the availability of a large and homogeneous cohort of patients, the goal of this study was to better characterize the biological behavior of this specific SFT subset located in the bone considering both the clinical, histological, and molecular features, as well as the applicability of the risk stratification model used for S-SFT.

2. Materials and Methods

The study was carried out on 24 patients affected by primary B-SFT treated at the Istituto Ortopedico Rizzoli between 1970 and 2019.

All patients were investigated, excluding history of meningeal SFT, whose metastatic bone localization could be misdiagnosed with a primary B-SFT. All cases were retrieved both from a radiological and clinical point of view through a review of medical records (anatomical site, tumor size, type of treatment, and surgical margins in the operated patients) and defined with regard to both the immuno-histochemical profile (positivity for CD34 and/or STAT6) and the molecular one (presence of the *NAB2-STAT6* gene fusion products, of the C228T and C250T *TERT* promoter variants, and of mutations in the p53 gene). The Demicco model [22] was used for the patient's risk stratification and the tumor size was assessed using the largest tumor dimension as a reference. All procedures were performed in accordance with the ethical standards of the Helsinki declaration. The study was approved by the ethical institutional committee on 22 July 2020 (study code: AVEC 730/2020/Oss/IOR). All analyses were completed with the help of the Statistical Package for Social Science (IBM Corp. Released 2013. IBM SPSS Statistics for Windows, Version 22.0 Armonk, NY, USA: IBM Corp.).

2.1. Histopathology Evaluation

Hematoxylin–eosin slides of all cases were reviewed by four pathologists (A.R., M.G., M.S., A.D.T.), and the morphological diagnosis of SFT was confirmed. Tumors were scored for mitotic figures, cellularity, nuclear pleomorphism, and presence of necrosis as universal and standardized criteria defining malignancy [17,18,27]. Mitotic index was calculated per 10 high-power fields (HPFs). The presence of high cellularity areas, defined as a hypercellular tumor with areas of nuclear overlap, and the presence of high pleomorphism, determined by hyperchromatic nuclei with foci of marked pleomorphism and bizarre cells according to Demicco criteria [22], were evaluated. Necrosis was scored as absence or minimal (<10%) or positive ($\geq$10%), based on available histological sections.

2.2. Immunohistochemistry

All paraffin-embedded tumor samples were evaluated by immunohistochemistry, as previously reported [27], with the following antibodies: CD34 (QBEnd-10; Ventana Medical Systems) and STAT6 (S-20, SC-621; Santa Cruz Biotechnology, Inc., Dallas, TX, USA). First, 4-μm-thick tissue sections were cut, heated at 58 $^\circ$C for 2 h, deparaffinized, and immunostained on a Ventana BenchMark following the manufacturer's guidelines (Ventana Medical Systems, Tucson, AZ, USA). The detection was performed using the UltraView Universal Alkaline Phosphatase Red Detection Kit and the UltraView Universal DAB Detection Kit (Ventana Medical Systems).

2.3. DNA and RNA Isolation

Sixteen tumor samples were available for molecular analyses. DNA and RNA were isolated from 10 formalin-fixed paraffin-embedded (FFPE) and 6 frozen tissues by using a QIAamp DNA FFPE Tissue Kit (Qiagen, Valencia, CA, USA) and RNeasy FFPE Tissue Kit (Qiagen), DNAzol and TRIzol (Invitrogen, Carlsbad, CA, USA) in accordance with the manufacturer's instructions.

2.4. Detection of NAB2-STAT6 Fusion Variants

The 24 most frequent *NAB2–STAT6* fusion variants found in S-SFT [27,29,30] were analyzed. PCR was performed by an AmplTaq Gold 360 Master Mix (Invitrogen) using 2 μL of cDNA product as previously described [27]. PCR products were sequenced using the BigDye Terminator v3.1 Cycle Sequencing Kit (Applied Biosystems) on an automated sequencer (ABI PRISM 3100 Genetic Analyzer 3130xl, Applied Biosystems, Foster City, CA, USA). To confirm the presence of specific *NAB2–STAT6* fusion breakpoints, the sequences were aligned using the CodonCode Aligner software (https://www.codoncode.com/aligner/, accessed on 29 April 2021).

2.5. TERT Promoter Mutation Analysis

The presence of C228T and C250T mutations at the *TERT* promoter region was primarily evaluated by Sanger sequencing as previously described [27].

Due to the low sensitivity of Sanger sequencing in detecting somatic mutations, we analyzed the same samples by Digital PCR (QuantStudio 3D Digital PCR System, Thermo Fisher Scientific, Waltham, MA, USA), used for rare allele detection to exclude the presence of *TERT* variants at low frequencies. We selected two TaqMan$^\circledR$ probe-based assays (Hs000000093_rm, Hs000000092_rm). Polymerase chain reaction amplification was carried out on a ProFlex^{TM} 2 $\times$ flat PCR System (Thermo Fisher Scientific, Waltham, MA, USA). Subsequent analysis and post-processing were performed by the QuantStudio^{TM} 3D AnalysisSuite^{TM}.

2.6. Analysis of p53 Mutation

To evaluate the presence of TP53 mutations, the samples were genotyped by direct sequence of all coding exons (2–11), including flanking intron-exon junctions. Sanger sequencing was performed using the BigDye Terminator v3.1 Cycle Sequencing Kit (Thermo Fisher Scientific) and the ABI PRISM 3500xL Geentic Analyzer (Thermo Fisher Scientific). To evaluate the presence of potential somatic copy number variations, a Microfluidic Chip-Based Digital PCR reaction was performed using the QuantStudio™3D Digital PCR System (QS3D, Thermo Fisher Scientific—US). A Taqman copy number assay was selected (Hs06423639_cn) to cover approximately the central part of the gene within exon 4 (location: hg38, Chr.17:7668402-7687550). RNase P gene was chosen as a reference locus. Polymerase chain reaction amplification was carried out on a ProFlex ™ 2 × Flat PCR System (Applied Biosystems). The fluorescence data were read and analyzed using QuantStudio 3D Analysis Suite Cloud Software. Results are expressed as copies per microliter and compared as a ratio of target (FAM)/Total (FAM + VIC) expressed in percentage. In case of a regular biallelic status, we expect this value to be around 50%. The TaqMan Copy Number probe was previously tested and validated on 6 DNA with a regular biallelic status of the p53 gene (with a target/total percentage range of 48.484–50.357%), confirming the absence of CNV alterations.

2.7. Statistics

Correlations between clinical, pathological, immuno-histochemical, and molecular data were assessed using contingency tables and chi-square test. The Kaplan–Meier method was used to estimate disease-specific survival (DSS), recurrence-free survival (RFS), and metastasis-free survival (MFS) based on histopathological criteria and the presence of *NAB2/STAT6* fusion variants.

MFS and DSS intervals were defined as the time between surgery and the first metastasis and death, respectively, or last follow-up available. Patients who died of other causes were excluded. By the log-rank test, differences in survival rates were assessed, considering p values < 0.05 as significant. For all analysis, was used the Statistical Package for Social Science (IBM Corp. Released 2013. IBM SPSS Statistics for Windows, Version 22.0. Armonk, NY, USA: IBM Corp.).

3. Results

3.1. Clinicopathological Evaluation

The clinical and pathological features of the 24 patients included in this study are summarized in Tables 1 and 2. The cohort is composed of 14 females (58.3%) and 10 males (41.7%) ranging from 7 to 84 years (mean 51 years). Most tumors arose in the axial skeleton (4 sacrum, 4 pubis, 2 scapula, and one lumbar vertebra), 9 in the lower extremities (6 femur, 2 tibia, and one fibula), and 4 in the upper extremities (humerus). In 5 out of 24 cases who were not feasible for surgery, only a biopsy was performed, followed by radiation therapy in two cases, chemotherapy in one, association of chemo- and radiation therapy in one, and embolization in one case. Nineteen patients underwent segmental resection or amputation with wide/radical margins in 16, intralesional margins in 2, and with marginal margins in one patient.

Table 1. Clinical features of 24 patients with a diagnosis of primary SFT of the bone.

Patient	Age	Sex	Anatomical Site	Tumor Size (cm)	Metastases	Surgical Procedure	Surgical Margins	DeMicco Score	Local Recurrence	Follow-Up (Months)	Status
1	75	F	Proximal tibia	15	Yes (3)	Thigh amputation	Wide	High	No	8	DOD
2	61	M	Distal femur	12.5	Yes (5)	Thigh amputation	Wide	High	No	15	DOD
3	66	M	Distal fibula	14	Yes (0)	Leg Amputation	Radical	Intermediate	No	58	DOD
4	71	F	Scapula	8	Yes (0)	Resection	Wide	Intermediate	No	28	DOD
5	84	M	Iliac wing	20	No	Inoperable	NA	Intermediate Inter	Yes (2)	2	DOD
6	52	F	Ileum pubic branch	7	No	Inoperable	NA	Low	Yes (24)	56	DOD
7	54	F	Femoral shaft	11	Yes (108)	Resection	Wide	Intermediate	Yes (149)	168	DOD
8	58	M	Sacrum	7.5	Yes (75)	Curettage	Marginal	Intermediate	Yes (97)	122	DOD
9	44	F	Distal femur	8.5	Yes (66)	Resection	Wide	Low	No	87	DOD
10	39	F	Proximal tibia	5	No	Resection	Wide	Low	No	421	NED
11	47	M	Scapula	6	No	Scapulectomy	Wide	Intermediate Intermediate	No	96	DOC
12	44	M	Sacrum	7	No	Inoperable	NA	Low	No	73	DOD
13	50	F	Proximal humerus	16	Yes (30)	ISTA	Radical	Intermediate	No	361	NED
14	58	M	Ischium pubic branch	8	No	Resection	Wide	Low	No	0	DOC
15	61	F	Iliac wing	8	Yes (84)	Resection	Intralesional	Low	No	137	DOD
16	27	F	Distal femur	17	No	IIAA	Intralesional	Intermediate	No	495	NED
17	32	F	Sacrum	15	No	Inoperable	NA	High	No	41	DOD
18	54	M	Sacrum	10	No	Resection	Wide	Low	No	67	NED
19	49	M	Proximal humerus	19	No	ISTA	Radical	High	No	92	DOC
20	31	F	4th lumbar vertebra	8	Yes (38)	Vertebrectomy	Wide	Low	No	66	DOD
21	21	F	Proximal humerus	9.5	Yes (72)	Resection	Wide	Intermediate	Yes (72)	72	AWD
22	7	F	Humeral shaft	11	No	Inoperable	NA	Intermediate	No	196	NED
23	60	F	Proximal femur	7.3	Yes (0)	Resection	Wide	Intermediate	No	21	DOD
24	69	M	Femoral shaft	10.5	No	Resection	Wide	High	No	18	NED

Legend: F: female; M: male; ISTA, interscapulothoracic amputation; IIAA, interileoabdominal amputation; NA, not applicable; AWD, Alive with disease; NED, non-evidence of disease; DOD, dead of disease; DOC, dead of other causes.

Table 2. Histopathological and molecular features of SFT patients.

Patient	Mitotic Count (X10 HPF)	Necrosis (≥10%)	STAT6	CD34	*TERT* Mutation	*NAB2-STAT6* Fusion	P53 Variants (Variant Type)	Grade of Malignancy
1	≥4	Yes	Pos	Pos	C250C/C228C	NA	yes (nonsense)	HIGH
2	≥4	Yes	Pos	Pos	C250C/C228C	EX6-EX17	yes (CNV deletion)	HIGH
3	<4	Yes	Pos	Pos	NA	NA	NA	LOW
4	≥4	Yes	Pos	Pos	C250C/C228C	NA	NA	HIGH
5	<4	No	Pos	Pos	NA	NA	NA	LOW
6	<4	No	Pos	Pos	NA	NA	NA	LOW
7	≥4	Yes	Pos	Pos	C250C/C228C	EX2-EX2	yes (CNV deletion)	HIGH
8	≥4	Yes	Pos	Pos	C250C/C228C	EX6-EX16	NA	HIGH
9	<4	No	Pos	Neg	NA	NA	NA	LOW
10	≥4	No	Pos	Pos	NA	NA	NA	HIGH
11	≥4	Yes	Pos	Pos	C250C/C228C	NA	yes (splice site)	HIGH
12	<4	No	Pos	Pos	C250C/C228C	EX4-EX2	NA	LOW
13	≥4	No	Pos	Pos	C250C/C228C	NA	yes (CNV deletion)	HIGH
14	<4	No	Pos	Pos	C250C/C228T	OTHER	no	LOW
15	<4	No	Pos	Pos	C250C/C228C	OTHER	yes (CNV deletion)	LOW
16	<4	Yes	Pos	Pos	C250C/C228C	EX6-EX16/EX6-EX17	yes (missense + CNV amplification)	LOW
17	≥4	Yes	Pos	Pos	NA	NA	NA	HIGH
18	<4	No	Pos	Pos	C250C/C228C	EX4-EX2	yes (CNV deletion)	LOW
19	≥4	Yes	Pos	Pos	C250C/C228C	EX6-EX17	yes (CNV deletion)	HIGH
20	≥4	No	Pos	Pos	NA	NA	NA	HIGH
21	≥4	Yes	Pos	Pos	C250C/C228C	EX6-EX17	yes (CNV duplication)	HIGH
22	≥4	No	Pos	Pos	NA	NA	NA	HIGH
23	≥4	Yes	Pos	Pos	C250C/C228C	EX6-EX17	yes (CNV deletion)	HIGH
24	≥4	Yes	Pos	Pos	C250C/C228C	EX4-EX2	NA	HIGH

Legend: HPF: high power fields; Pos, positive; Neg, negative; NA, not applicable; MAL, malignant.

In the group of 19 patients surgically treated, 3 patients (16%) developed local recurrence at a mean time of 106 months (range 72–149 months, median 97 months).

Twelve patients (50%) had metastasis (9 localized at lungs and 3 to bone): 3 patients had metastasis at presentation (in one case, lung; in one case, soft tissues; and in one case both lungs and bone): chemotherapy (CT) with a combination of doxorubicin, methotrexate cisplatin, and ifosfamide was given to the two patients with lung metastasis at presentation. The other 9 (37.5%) patients developed metastasis at a mean time of 53 months (range 3–108 months).

Nine patients with localized disease received chemotherapy with a combination of doxorubicin, methotrexate, cisplatin, and ifosfamide (eight adjuvant and one neoadjuvant chemotherapy) whereas four patients underwent radiation therapy (two adjuvant radiation 185 and two for palliation). One patient underwent selective arterial embolization with palliative intent.

The mean follow-up was 112 months (0–495), median 69 months. At the last follow-up, 14 patients out of 24 were dead of disease (DOD), 3 dead of other causes (DOC), one alive with metastatic disease (AWD), and 6 alive without evidence of disease (NED). Radiologically, all cases were lytic, with areas of sclerosis in two cases. Mean tumor size was 10.87 cm (range 5–20 cm); in 13 cases, it was $\leq$10cm while in the other 11 cases, it was >10 cm (Table 1).

3.2. Histopathological and Immunohistochemical Features

From a histopathological point of view, 15 cases showed more than 4 mitotic figures per 10 HPF and were associated with high cellularity, cytologic atypia, and >10% of necrosis, defining high-grade tumors (Figure 1, Table 2). CD34 and STAT6 immunopositivity was observed in 95% (23/24) and in 100% (24/24) of cases, respectively (Figure 2).

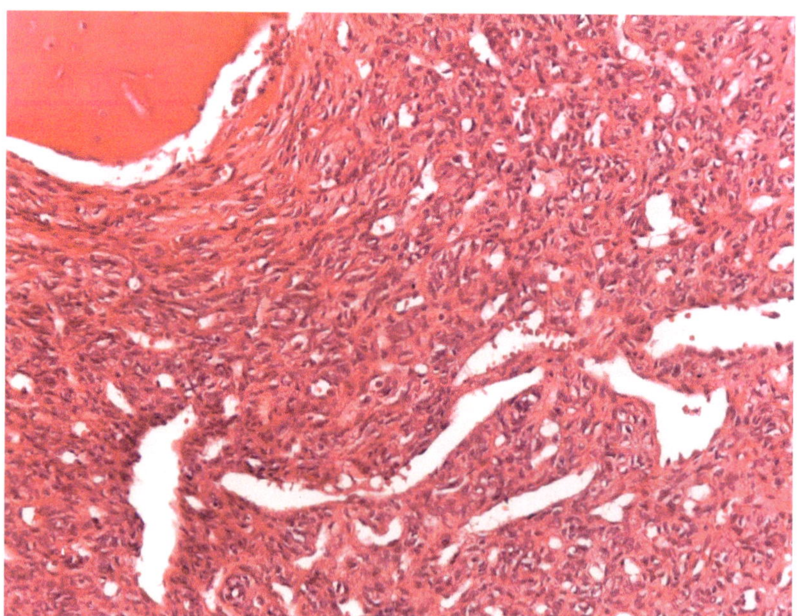

Figure 1. Solitary fibrous tumor: A spindle cell proliferation showing hemangiopericytoma-like blood vessels is seen (Hematoxylin &Eosin, original magnification, ×100).

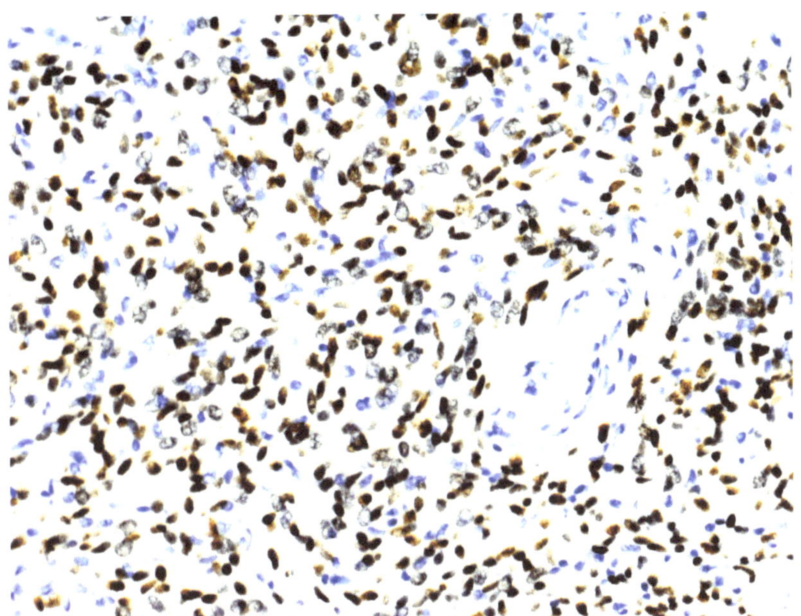

Figure 2. The nuclei of neoplastic cells express STAT6 (original magnification, ×200).

According to Demicco score [22], 8 patients (33%) were classified in the low-risk group, 11 (46%) in the intermediate-risk group, and 5 (21%) in the high-risk group (Table 1).

Two of the nine patients who developed distant metastasis belonged to the low-risk groups, five to the intermediate-risk group, while two patients belonged to the high-risk group. The three patients with metastasis at presentation were equally distributed in the three risk groups.

3.3. NAB2–STAT6 Fusion Variants

The analysis of fusion transcripts identified *NAB2–STAT6* fusion variants in 10 out of 12 (83.3%) samples (Table 2). In two cases, no variant was found. Considering the 24 types of fusion variants evaluated, 2 breakpoints were detected with a higher frequency: *NAB2exon6—STAT6exon17* (4 cases) and *NAB2exon4-STAT6exon2* (3 cases), followed by the breakpoint *NAB2exon6—STAT6exon16, NAB2exon2—STAT6exon2* and *NAB2exon6-STAT6exon16/NAB2exon6—STAT6exon17* in one case (Table 2). Regarding the Demicco score risk, the *NABex6-STAT6ex17* fusion variant was present only in high- and intermediate-risk patients, even if *NAB2-STAT6* fusion variants and Demicco score risk were not significantly correlated ($p = 0.25$).

3.4. TERT Promoter Mutations: C228T and C250T

The wild-type C250C genotype was shown in all 16 samples while no C250T mutations were detected. In only one DNA sample, a heterozygous C228T substitution was detected.

The only patient presenting this variant died one day after surgery due to complications; therefore, it was not possible to evaluate its prognostic role (Table 2).

3.5. p53 Mutations

Overall, we detected p53 genetic alterations in 11 samples (Table 2). Three samples presented point mutations: a nonsense heterozygous variant (p.Gln165*) was detected in patient 1; a missense heterozygous variants (p.Ala63Val), already described as a variant of uncertain significance (VUS), was detected in patient 16; and a homozygous splice site

alteration (c.375 + 1G > A) was observed in patient 11. All samples except two (1 and 14
showed the presence of a copy number variation (CNV) involving at least exon 4 of *p53*
In detail, CNV deletions were detected in patient 2, 7, 13, 15, 18, 19, and 23 whereas CNV
amplifications were detected in patient 16 and 21.

3.6. Correlations between Clinicopathological, Immunohistochemical, and Molecular Data

Regarding the entire population of study (24 cases), 5- and 10-year DSS were respec
tively 64% and 42%, whereas on the localized tumor, 5- and 10-year disease-related-specific
DSS were respectively 80% and 60%. As expected, localized and surgically treated pa
tients (16 out of 24, 66%) showed a better 5-year DSS than metastatic ones (74% vs. 33%
(Figure 3).

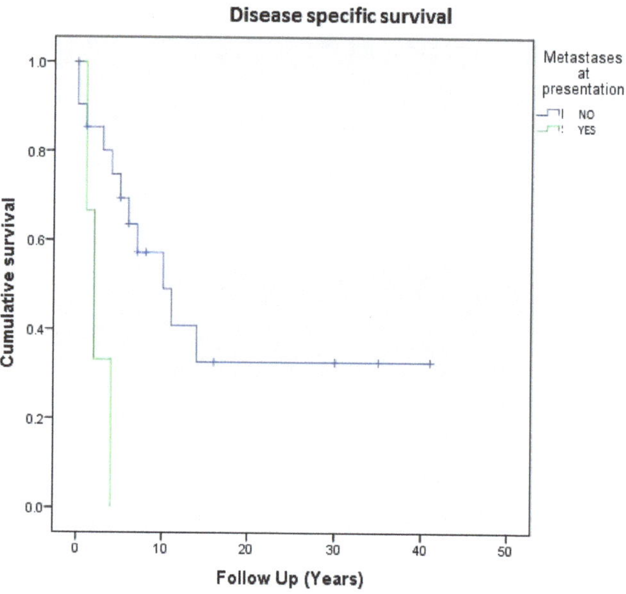

Figure 3. Cumulative survival of 24 patients affected by primary bone solitary fibrous tumor
stratification by metastases at presentation.

Table 3 summarizes the results of the Kaplan–Meier survival analysis of the clinicopathological variables (histological grade, tumor size, age, mitosis, necrosis, Demicco
score risk). Stratification by tumor size did not correlate with DSS either for localized
patients ($p = 0.54$) or for the whole series ($p = 0.44$). However, the only patient with tumor
size <5 cm was alive at follow up (Table 3). Stratification based on mitotic count was carried
out (A $\leq$ 1 mitosis, B = 1–3 mitosis, and C $\geq$ 4); no correlation was found in terms of DSS
at the 5- and 10-year follow up either for the whole series ($p = 0.54$) or for patients with
localized disease ($p = 0.33$) (Table 3).

No significant differences in terms of DSS were found between the different variables
analyzed by univariate analysis. Of interest, DSS in patients aged <55 and $\geq$55 years was
found to be almost near statistical significance ($p = 0.06$), confirming a better prognosis in
younger patients. In line with the malignancy histological criteria, none of the *NAB2-STAT6*
fusion variants detected were significantly correlated to DSS both in all 24 cases ($p = 0.72$)
and in 16 localized cases ($p = 0.57$). In localized patients, Exon6 was involved in 2 cases out
of 5 while other fusion variants (Exon2, Exon4, Other) were detectable in 3 cases out of 6; no
significant correlation ($p = 0.68$) in terms of DSS was observed at the 5- and 10-year follow
up (80% vs. 40% and 100% vs. 67%, respectively). *P53* variants were assessed in 12 cases:
11 (91.6%) cases showed variation while in one case, no alteration was found. Since few

cases were analyzed, no statistical analysis was done; however, tumors with p53 mutations were classified as follows: two 'low-risk', three 'high-risk', and six 'intermediate-risk' cases. Further, 5- and 10-year DSS in the mutated patient was 73% and 54%, respectively, with a mean follow up of 139 months (range 8–495).

Table 3. Disease-specific survival (DSS) analysis related to clinicopathological parameters.

Variables	Disease Specific Survival (24 pts)			Localized Disease * (16 pts)		
	5 Years-DSS	10 Years-DSS	*p*-Value	5 Years-DSS	10 Years-DSS	*p*-Value
Histological Grade						
Low	62%	31%	0.52	100%	67%	0.84
High	65%	58%		82%	71%	
Size						
(A) 0–4.99 cm	100%	100%	0.44	100%	100%	0.54
(B) 5–9.99 cm	70%	36%		87%	45%	
(C)10–14.99 cm	62%	62%		80%	80%	
(D) >15	50%	50%		50%	50%	
Age						
<55 years	86%	27%	0.06	100%	77%	0.15
≥55 years	61%	27%		100%	60%	
Mitosis						
(A) <1	60%	30%	0.54	60%	30%	0.33
(B) 1–3	66%	33%		100%	50%	
(C) ≥4	65%	58%		76%	68%	
Necrosis						
<10%	80%	47%	0.66	100%	62%	0.95
≥10%	51%	51%		78%	78%	
Gene Fusion						
Exon6				80%	40%	0.68
Other				100%	67%	
Demicco Score Risk						
High				54%	54%	0.43
Intermediate				72%	46%	
Low				64%	28%	

* Surgically treated patients; DSS: disease-specific survival.

The MFS was found to be about 72% at 5 years and 27% at 10 years, as 9 out of 16 patients developed distant metastasis after a mean time of 53 months, whereas the RFS was found to be 100% at 5 years and 75% at 10 years, respectively, as 3 patients out of 16 developed local recurrence after a mean time of 106 months. No significant differences in terms of MFS and of RFS were found between the different variables analyzed by univariate analysis.

Of interest was finding that no local recurrence occurred in patients considered to be low-grade malignancy. In particular, 10-year RFS was 64% for high-grade patients against 100% in low-grade patients. Nevertheless, the *p* value obtained was not significant (*p* = 0.19), probably due to the limited number of patients, which could represent a bias.

4. Discussion

Primary B-SFT represents an extremely rare entity and to date any correlations between histopathological, immuno-histochemical, and molecular features and DSS have not yet been determined due to the lack of sufficiently numerous cases reported. Despite the rarity of this pathology, this is extremely important in order to stratify patients in terms of risk of relapse and distant metastasis and thus define the best treatment and surveillance strategies. To the best of our knowledge, the present study reports the clinical, histopathological, and molecular characteristics of the largest series reported in the literature. The primary aim of this work was to find any clinical and molecular prognostic factors and to compare them with those currently used for S-SFT [27], evaluating the possibility of a different behavior between SFT originating from the bone and from soft tissue, even if they share the same histology. Secondly, we applied the risk stratification system proposed by Demicco et al. in 2017 [22] to our selected series of 16 patients with resectable and localized primary B- SFT at onset, in order to understand whether this system, already evaluated by us previously on patients affected by S- SFT of the extremities, was able to correctly predict the risk of local and distant metastatic relapse even in the case B-SFT.

Despite the fact that SFTs of the bone and of soft tissue share the same morphological features, the data obtained in this series of the B-SFT did not confirm those already obtained by us on the S-SFT series, comparable to those available for other cases reported in the literature [14,15,18,22,24,29].

In particular, no correlations emerged between DSS, RFS, and MFS with clinicopathological variables (histological grade, tumor size, age, mitosis, necrosis, Demicco score risk), unlike what was reported by Gold and Barthelmess [14,25], and molecular features (*TERT* promoter mutations [14,21,25,29] and *NAB2-STAT6* fusion transcripts variants). These last results appear to be in line with those reported by Machado and Bianchi [21,27]. Data of interest was the absence of local recurrence in all low-risk patients (according to Demicco scoring system), although without evidence of statistical correlation.

TERT promoter mutation in a heterozygous state (C250C/C228T) was only found in one case out of 16. Unfortunately, the patient died the day after surgery due to complications, preventing assessment of the possibility of any correlations with this mutation. These data differ markedly from those obtained by Gold, Machado, and Barthelmess [14,21,25] and from those in the case series of S-SFT of the extremities presented by Bianchi and collaborators, in which the frequency of mutations of the promoter of *TERT* was found to be nearly 50% and 23.7%, respectively. In particular, all three metastatic patients of our previous study presented C228T site mutation in a homozygous state [29].

Considering p53, almost all evaluated samples (91.6%) showed a genetic variant, different from what has been reported in the literature [21]. Despite the limited number of samples, 9 of the 11 tumors with p53 mutations were classified as 'high' or 'intermediate', thus confirming results detailed in a previous study [21]. Further studies will be required to evaluate the inclusion of p53 genetic status in the risk stratification system. Regarding *NAB2-STAT6* fusion transcript variants, the most frequently encountered was *NAB2ex6-STAT6ex17* (4 out of 12 cases), followed by *NAB2ex4-STAT6ex2* in 3 cases, *NAB2ex2-STAT6ex2* in one case, *NAB2ex6-STAT6ex16/NAB2ex6-STAT6ex17* in one case, and *NAB2ex6-STAT6ex16* in another. In two cases, there was evidence of different breakpoints from the 24 most frequently evaluated, which were therefore encoded as the other. In contrast, in our S-SFT series, the most frequently reported variants were *NAB2ex6-STAT6ex17*, *NAB2ex6-STAT6ex2*, and *NAB2ex4-STAT6ex4* [27]. In both series, no statistically significant correlations emerged between the different fusion variants and the oncological outcome, different from Barthelmess and Tai, who reported a better prognosis for *NAB2ex4-STAT6ex2/4* variants associated with a lower mitotic count and relapse rate [25,29]. However, there was a tendency for the *NAB2ex6-STAT6ex17* fusion variants to be associated with high- and intermediate-risk neoplasms according to the Demicco system [22]; however, this did not result in statistically significant values ($p = 0.25$).

Among the limitations of this study, first, the population under study was small due to the extreme rarity of this pathology. Furthermore, there was a lack of uniformity regarding the type of treatment of the cases treated, because of the long period evaluated, during which the therapeutic approaches changed: most of the non-operated patients date back to the first decades of this period, while over the years, surgical treatments have gradually become more and more conservative and have been associated with adjuvant chemotherapy, and adjuvant or palliative radiation therapy in some cases. It is important to underline that ancillary genetic investigations, such as FISH and RT-PCR, are not yet of practical use in molecular diagnostics and are not always feasible in all cases because of the decalcification process that occurs on bone samples.

5. Conclusions

In conclusion, no correlation emerged between Demicco's risk assessment criteria and clinical behavior as evidenced for the S-SFT. In fact, the clinicopathological criteria of malignancy devised for SFT of soft tissues failed to predict outcomes in primary SFT of the bone. Further validation on more numerous as well as more homogeneous samples is necessary to validate some molecular differences between primitive SFT of the bone with respect to that of soft tissues and to evaluate the eventual prognostic implications.

Author Contributions: Conceptualization, A.P.D.T. and M.G.; methodology, E.P., L.P., D.L., G.B. and M.S.; software, G.B.; validation, A.R.; resources, K.S., C.F., L.S. and M.G.; data curation, I.B., D.L., G.B. and A.R.; writing—original draft preparation, D.L. and G.B.; writing—review and editing, A.R., L.P. and E.P. All authors have read and agreed to the published version of the manuscript.

Funding: This research received no external funding.

Institutional Review Board Statement: The study was conducted according to the guidelines of the Declaration of Helsinki, and approved by the Ethics Committee of Area Vasta Emilia Centro della Regione Emilia-Romagna (protocol code: AVEC 730/2020/Oss/IOR and date of approval 22 July 2020).

Informed Consent Statement: Informed consent was obtained from all subjects involved in the study.

Data Availability Statement: Data available on request due to restrictions e.g., privacy or ethical. The data presented in this study are available on request from the corresponding author.

Acknowledgments: The authors are grateful to Muscolo Skeletal Tumor Biobank-Biobanca dei Tumori Muscoloscheletrici (Biotum)—member of the CRB-IOR—which provided us the biological samples. This work was supported by the SELNET-ID 825806 Project-PI Scotlandi Katia and Ministero della Salute (5 × 1000, Anno 2018 Redditi 2017, contributions to the IRCCS, Istituto Ortopedico Rizzoli).

Conflicts of Interest: The authors declare no conflict of interest. The funders had no role in the design of the study; in the collection, analyses, or interpretation of data; in the writing of the manuscript, or in the decision to publish the results.

References

1. Demicco, E.G.; Fritchie, K.J.; Han, A. Solitary fibrous tumor. In *WHO Classification of Tumours 5th Edition: Soft Tissue and Bone Tumours*; WHO Classification of Tumours Editorial Board, Ed.; IARC Press: Lyon, France, 2020; Volume 5, pp. 104–108.
2. Klemperer, P.; Rabin, C.B. Primary neoplasms of the pleura. *Arch. Pathol.* **1931**, *11*, 385–412.
3. Picci, P.; Manfrini, M.; Donati, D.M.; Gambarotti, M.; Righi, A.; Vanel, D.; Dei Tos, A.P. *Diagnosis of Musculoskeletal Tumors and Tumor-Like Conditions. Clinical, Radiological and Histological Correlations—The Rizzoli Case Archive*, 2nd ed.; Springer: Cham, Switzerland, 2020; pp. 3–11.
4. Verbeke, S.L.; Fletcher, C.D.; Alberghini, M.; Daugaard, S.; Flanagan, A.M.; Parratt, T.; Kroon, H.M.; Hogendoorn, P.; Bovée, J.V. A Reappraisal of Hemangiopericytoma of Bone; Analysis of Cases Reclassified as Synovial Sarcoma and Solitary Fibrous Tumor of Bone. *Am. J. Surg. Pathol.* **2010**, *34*, 777–783. [CrossRef]
5. Dei Tos, A.P.; Righi, A.; Gambarotti, M.; Vanel, D.; Ferrari, C.; Benini, S.; Ferrari, S.; Picci, P. Reappraisal of primary spindle/pleomorphic sarcoma of bone. *Mod. Pathol.* **2014**, *27*, 15A.
6. Fletcher, C.D. Haemangiopericytoma- A dying breed? Reappraisal of an 'entity' and its variants: A hypothesis. *Curr. Diagn. Pathol.* **1994**, *1*, 19–23. [CrossRef]

7.   Fletcher, C.D. The evolving classification of soft tissue tumours: An update based on the new WHO classification. *Histopathology* **2006**, *48*, 3–12. [CrossRef]

8.   Coca-Pelaz, A.; Llorente-Pendás, J.L.; Vivanco-Allende, B.; Suarez-Nieto, C. Solitary fibrous tumor of the petrous bone: A successful treatment option. *Acta Oto-Laryngol.* **2011**, *131*, 1349–1352. [CrossRef]

9.   Ge, X.; Liao, J.; Choo, R.J.; Yan, J.; Zhang, J. Solitary fibrous tumor of the ilium. *Medicine* **2017**, *96*, e9355. [CrossRef]

10.   Suarez-Zamora, D.A.; Rodriguez-Urrego, P.A.; Soto-Montoya, C.; Rivero-Rapalino, O.; Palau-Lazaro, M.A. Malignant Solitary Fibrous Tumor of the Humerus: A Case Report of an Extremely Rare Primary Bone Tumor. *Int. J. Surg. Pathol.* **2018**, *26*, 772–776. [CrossRef] [PubMed]

11.   Jia, C.; Crim, J.; Evenski, A.; Layfield, L.J. Solitary fibrous tumor of bone developing lung metastases on long-term follow-up. *Skelet. Radiol.* **2020**, *49*, 1865–1871. [CrossRef] [PubMed]

12.   Sbaraglia, M.; Righi, A.; Gambarotti, M.; Vanel, D.; Picci, P.; Tos, A.P.D. Soft Tissue Tumors Rarely Presenting Primary in Bone: Diagnostic Pitfalls. *Surg. Pathol. Clin.* **2017**, *10*, 705–730. [CrossRef] [PubMed]

13.   England, D.M.; Hochholzer, L.; McCarthy, M.J. Localized Benign and Malignant Fibrous Tumors of the Pleura. *Am. J. Surg. Pathol.* **1989**, *13*, 640–658. [CrossRef] [PubMed]

14.   Gold, J.S.; Antonescu, C.R.; Hajdu, C.; Ferrone, C.R.; Hussain, M.; Lewis, J.J.; Brennan, M.F.; Coit, D.G. Clinicopathologic correlates of solitary fibrous tumors. *Cancer* **2002**, *94*, 1057–1068. [CrossRef] [PubMed]

15.   Gholami, S.; Cassidy, M.R.; Kirane, A.; Kuk, D.; Zanchelli, B.; Antonescu, C.R.; Singer, S.; Brennan, M. Size and Location are the Most Important Risk Factors for Malignant Behavior in Resected Solitary Fibrous Tumors. *Ann. Surg. Oncol.* **2017**, *24*, 3865–3871. [CrossRef] [PubMed]

16.   Kim, J.M.; Choi, Y.-L.; Kim, Y.J.; Park, H.K.; Kima, J.M.; Choia, Y.-L.; Parka, H.K. Comparison and evaluation of risk factors for meningeal, pleural, and extrapleural solitary fibrous tumors: A clinicopathological study of 92 cases confirmed by STAT6 immunohistochemical staining. *Pathol. Res. Pract.* **2017**, *213*, 619–625. [CrossRef]

17.   Doyle, L.A.; Fletcher, C.D.M. Predicting behavior of solitary fibrous tumor: Are we getting closer to more accurate risk assessment? *Ann. Surg. Oncol.* **2013**, *20*, 4055–4056. [CrossRef]

18.   Demicco, E.G.; Park, M.S.; Araujo, D.M.; Fox, P.S.; Bassett, R.L.; E Pollock, R.; Lazar, A.J.; Wang, W.-L. Solitary fibrous tumor: A clinicopathological study of 110 cases and proposed risk assessment model. *Mod. Pathol.* **2012**, *25*, 1298–1306. [CrossRef]

19.   Pasquali, S.; Gronchi, A.; Strauss, D.; Bonvalot, S.; Jeys, L.; Stacchiotti, S.; Hayes, A.; Honore, C.; Collini, P.; Renne, S.L.; et al. Resectable extra-pleural and extra-meningeal solitary fibrous tumours: A multi-centre prognostic study. *Eur. J. Surg. Oncol. (EJSO)* **2016**, *42*, 1064–1070. [CrossRef]

20.   Salas, S.; Resseguier, N.; Blay, J.Y.; Le Cesne, A.; Italiano, A.; Chevreau, C.; Rosset, P.; Isambert, N.; Soulie, P.; Cupissol, D.; et al. Prediction of local and metastatic recurrence in solitary fibrous tumor: Construction of a risk calculator in a multicenter cohort from the French Sarcoma Group (FSG) database. *Ann. Oncol.* **2017**, *28*, 1779–1787. [CrossRef]

21.   Machado, I.; Morales, G.N.; Cruz, J.; Lavernia, J.; Giner, F.; Navarro, S.; Ferrandez, A.; Llombart-Bosch, A. Solitary fibrous tumor: A case series identifying pathological adverse factors—implications for risk stratification and classification. *Virchows Arch.* **2019**, *476*, 597–607. [CrossRef]

22.   Demicco, E.G.; Wagner, M.J.; Maki, R.G.; Gupta, V.; Iofin, I.; Lazar, A.J.; Wang, W.-L. Risk assessment in solitary fibrous tumors: Validation and refinement of a risk stratification model. *Mod. Pathol.* **2017**, *30*, 1433–1442. [CrossRef] [PubMed]

23.   Chmielecki, J.; Crago, A.M.; Rosenberg, M.; O'Connor, R.; Walker, S.R.; Ambrogio, L.; Auclair, D.; McKenna, A.; Heinrich, M.C.; Frank, D.A.; et al. Whole-exome sequencing identifies a recurrent NAB2-STAT6 fusion in solitary fibrous tumors. *Nat. Genet.* **2013**, *45*, 131–132. [CrossRef]

24.   Robinson, D.R.; Wu, Y.-M.; Kalyana-Sundaram, S.; Cao, X.; Lonigro, R.J.; Sung, Y.-S.; Chen, C.-L.; Zhang, L.; Wang, R.; Su, F.; et al. Identification of recurrent NAB2-STAT6 gene fusions in solitary fibrous tumor by integrative sequencing. *Nat. Genet.* **2013**, *45*, 180–185. [CrossRef]

25.   Barthelmeß, S.; Geddert, H.; Boltze, C.; Moskalev, E.A.; Bieg, M.; Sirbu, H.; Brors, B.; Wiemann, S.; Hartmann, A.; Agaimy, A.; et al. Solitary Fibrous Tumors/Hemangiopericytomas with Different Variants of the NAB2-STAT6 Gene Fusion Are Characterized by Specific Histomorphology and Distinct Clinicopathological Features. *Am. J. Pathol.* **2014**, *184*, 1209–1218. [CrossRef]

26.   Bahrami, A.; Lee, S.; Schaefer, I.-M.; Boland, J.M.; Patton, K.T.; Pounds, S.; Fletcher, C.D. TERT promoter mutations and prognosis in solitary fibrous tumor. *Mod. Pathol.* **2016**, *29*, 1511–1522. [CrossRef] [PubMed]

27.   Bianchi, G.; Sambri, A.; Pedrini, E.; Pazzaglia, L.; Sangiorgi, L.; Ruengwanichayakun, P.; Donati, D.; Benassi, M.S.; Righi, A. Histological and molecular features of solitary fibrous tumor of the extremities: Clinical correlation. *Virchows Arch.* **2020**, *476*, 445–454. [CrossRef]

28.   Park, H.K.; Yu, D.B.; Sung, M.; Oh, E.; Kim, M.; Song, J.-Y.; Lee, M.-S.; Jung, K.; Noh, K.-W.; An, S.; et al. Molecular changes in solitary fibrous tumor progression. *J. Mol. Med.* **2019**, *97*, 1413–1425. [CrossRef]

29.   Tai, H.-C.; Chuang, I.-C.; Chen, T.-C.; Li, C.-F.; Huang, S.-C.; Kao, Y.-C.; Lin, P.-C.; Tsai, J.-W.; Lan, J.; Yu, S.-C.; et al. NAB2–STAT6 fusion types account for clinicopathological variations in solitary fibrous tumors. *Mod. Pathol.* **2015**, *28*, 1324–1335. [CrossRef] [PubMed]

30.   Akaike, K.; Kurisaki-Arakawa, A.; Hara, K.; Suehara, Y.; Takagi, T.; Mitani, K.; Kaneko, K.; Yao, T.; Saito, T. Distinct clinicopathological features of NAB2-STAT6 fusion gene variants in solitary fibrous tumor with emphasis on the acquisition of highly malignant potential. *Hum. Pathol.* **2015**, *46*, 347–356. [CrossRef] [PubMed]

MDPI AG
Grosspeteranlage 5
4052 Basel
Switzerland
Tel.: +41 61 683 77 34

Cancers Editorial Office
E-mail: cancers@mdpi.com
www.mdpi.com/journal/cancers